Engraved Handguns of .22 Calibre

Engraved Handguns

of .22 calibre

1855-1885

Engraved Handguns of .22 Calibre

1855-1885

John S. Laidacker

Schiffer Military History
Atglen, PA

Dedication

To Hank Truslow, without whose encouragement
this project would never have been completed.

The border of the title page (p 2) is taken from an advertising broadside used by John Krider, Gun Maker, of Philadelphia, Pennsylvania. The broadside was produced in the mid-ninteenth century.

Book Design by Ian Robertson.

Library of Congress Catalog Number: 2002112029

Printed in China.
ISBN: 0-7643-1683-4

We are interested in hearing from authors with book ideas on military topics.

Published by Schiffer Publishing Ltd.
4880 Lower Valley Road
Atglen, PA 19310 USA
Phone: (610) 593-1777
FAX: (610) 593-2002
E-mail: info@schifferbooks.com.
Visit our web site at: www.schifferbooks.com
Please write for a free catalog.
This book may be purchased from the publisher.
Please include $3.95 postage.
Try your bookstore first.

In Europe, Schiffer books are distributed by:
Bushwood Books
6 Marksbury Road
Kew Gardens
Surrey TW9 4JF
England
Phone: 44 (0)181 392-8585
FAX: 44 (0)181 392-9876
E-mail: Bushwd@aol.com.

Try your bookstore first.

Contents

Introduction

For many years the designs found on engraved handguns has been found quite interesting. This is an attempt to identify motifs employed, and to determine engraving characteristics of various arms producing companies.

To accomplish this, examples have been examined and photographed, and a system of identification has been structured. By using this system, one may determine styles used within a particular company, or to compare similar styles used by more than one company.

This cannot be construed as a definitive work, nor is it near all-inclusive on the subject. It is simply the findings of one person's curiosity, based on a very limited number of examples, from which conclusions have been drawn. The study is only a crude beginning. Hopefully others, by pooling information, will eventually bring forth a more informative and comprehensive coverage.

Content is confined to handguns of .22 calibre, as produced from 1855 to the mid-1870s. It is a companion piece to the book "Collected Notes Concerning DEVELOPMENTAL CARTRIDGE HANDGUNS IN .22 CALIBRE As Produced in the United States and Abroad From 1855 to 1875." (John S. Laidacker, Library of Congress Catalog Card Number 93-91773; hereinafter referred to as "Notebook.")

This is a very small segment of firearm engraving throughout the years. To gain a wider perspective, one should study engraving patterns as employed on all firearms during all periods. A selected bibliography appears at the end of this study; the books listed have several illustrations depicting gun engraving patterns.

No basic data relative to differentiation of models or company backgrounds are included. If such be desired, one should consult book mentioned in preceding paragraph.

Questions arise:

Who did the engraving, and under what circumstances?

Was it performed on location where the guns were produced, or "farmed out" to independent engravers?

How is it determined whether the engraving was done by or for the gun producing company, or for an individual after the gun had left the plant?

Was the engraving done by or for the company, or was the engraving executed by a distributor and/or retailer?

These questions to a very large degree are unanswered. When something is known for a specific company, that information will be listed.

All examples illustrated herein are or were in my collection. Borrowing examples for inclusion from other collections would have expanded coverage, but by doing such, more time than is available would have been required. As previously mentioned, this constitutes a beginning, not an end.

Some entries concern handguns produced after the mid 1870s. Generally, as the quality of the guns themselves deteriorated after this date, so did the quality of engraving. More punchwork and more shallow engraving is observed on these pieces, and the motifs themselves were more crude and less artistic - enter the period of the "suicide special." This may be attributed to harsh financial times rather than lack of desire to create a superior product. Throughout this later period, the major companies still had fine engravings, but the great mass of guns produced were inferior in all respects to those made earlier.

As with other art forms, the quality of design and execution covers a wide range. Pieces of special order or of presentation grade naturally were of better quality. Standard engraving, available from stock by the larger makers, usually conform to a standard pattern.

Nimsehke, the Ulrichs, Wolf, and Young were well known master engravers. Even they, with a few exceptions for a few years, were not full time employees of one specific company. The others, which constitute the vast majority of pieces encountered, are anonymous. Probably firms utilized the skills of local engravers whose work was not confined to firearms alone. Those engravers who have been identified were from Germany, and used the patterns popular in that country during the time period covered by this work.

Two listings concerning price of engraving have been located:

The Enterprise Gun Works (James Bown & Son) of Pittsburgh, Pennsylvania, catalogue of 1876 lists prices for the Smith & Wesson Model One 3rd Variation revolver as follows: Blue finish $8; nickel Plate finish $9; engraved, with ivory stocks $14.50; engraved, with pearl stocks $16.00.

In the catalogue of the New York City firm of Homer Fisher, issued in 1879, engraving of .22 calibre revolvers was offered for 50¢. The firm also offered ivory handles for $1.00, and pearl handles for $1.50.

The Smith & Wesson offered by Bown probably had factory engraving (as no other maker's products were listed as engraved.)

Note the price difference between what is probably factory engraving compared to engraving by a retailer. In the 1870s, the average worker was fortunate if he could earn one dollar for a ten-hour workday.

Conclusion: By examining the engraving pictured within this work, one should develop an understanding of the main patterns of the period—and very little else. But, that is why it was done....

Format Explanation

The number in top right and left hand corners of each page is the **Description Number** for the first entry under each producer. If followed by a +, more than one model by that producer is included. It there be no +, a specific model is indicated. These numbers relate to description numbers listed in the book "Developmental Cartridge Handguns In .22 Calibre."

Example: 572+ = Smith & Wesson; 584+ = Whitneyville; 110 = Remington Elliot.

A number in **bold face** near a picture is the serial number of the piece illustrated.

The **Pattern Numbers** indicate a specific design appearing on one location on a piece by one producer.

Each manufacturer has a set of pattern numbers exclusive to that producer; one Pattern Number is not of the same design for every company.

The same design is often used by more than one maker. When so, comparisons are made by abbreviating design and cross-referencing thusly:

572+33	similar to	22+30
Smith & Wesson Pattern 33	similar to	Whitneyville Pattern 30

(30 series of numbers relates to engraving on side of frame.)

The Pattern Numbers are divided into following categories:

1 Barrel
30 Side of frame
40 Cylinder topstrap
60 Cylinder
70 Frame backstrap
90 Area forward of cylinder

Where sufficient comparative examples by one maker are available, a chart listing patterns used on each engraved area is present. Illustrations appear after chart sections under each company heading.

The **Miscellaneous Section** is composed of single entries where only one or a few examples are available for discussion. There are no Pattern Numbers assigned to these entries.

A **Pattern Explanation** section follows. This hopefully will familiarize the reader with terms used throughout this work.

Explanation of Terms Used in Describing Patterns

There are undoubtedly technical or more accepted terms for specific patterns or applications, but they are not known to writer. In lieu of such, names have been given to particular patterns as they appear to compiler's eye.

A few patterns were used on handguns of several companies during the period covered, roughly 1855 to 1885. Whoever the engravers were, they in general used many of the same patterns and methods of accenting these patterns.

The "American style" of engraving patterns is really a continuation, with some modifications, of the prevalent German patterns brought to this country by German engravers during the nineteenth century. Most of the engravers during the period being examined were of German extraction.

Illustrations of the major patterns follow, along with references to handgun upon which it is found.

The most popular design was the **foliate shaded C-scroll with stippled background.** Composed of curved scrolls resembling the letter C, with radiating leaves or petals. Lines within segments accentuate the area, referred to as shading. Clusters of small punched circles are referred to as **stippling.**

The example on this page is cataloged as 572+71 (meaning Description Number 572, which is Smith & Wesson Model 1 3rd Variation, pattern 71.) If one turns to the Smith & Wesson section (signified by 572+ in top corners of each page), one will read Pattern Description and note that example is serial number 45533.

All illustrations within this EXPLANATION will bear similar number combinations which will identify maker, pattern number, and specific pistol pictured.

572 = 71

Ellipse. Another very common pattern. Usually in series, many times itermingled with **diamonds** on barrels and/or backstraps. Sometimes referred to as "entwining lines." Often stamped hollow circles are applied at junctures or terminals of linking geometric patterns as seen below. Illustration has a double line ellipse followed by a single line ellipse.

522+71

Note that Pattern Numbers differ among makers. Preceding illustrations are each numbered 71, but are found on backstraps of pistols made by different makers. They are quite different, but each numbered 71 under each maker.

Quilted diamond. Series of connecting diamonds, usually quite small. These may be used as central motif within a larger figure, as background in lieu of **shading** or **stippling**, or to fill in very small secondary areas. More often than not, a small circle (hollow or solid) is punched into the centre of each diamond.

Quilted diamonds within two larger diamonds.

Note **petals** above and below the outer diamond. Also note series of shaded **scallops** bordering sides of backstrap.

The centers of diamonds are flat, not pointed as found in **checkering.**

522+74

105+75

Checkering. A diamond pattern wherein cuts are so deep that center of each diamond comes to a point. The angle of line intersection will vary from piece to piece; usually a diamond pattern is created, though may be in form of square.

Two sides of frames in inspection plate area are now pictured, each having a variation of the **Foliate C-scroll** pattern.

Primary scroll and extended leaves begin at base and move upward.

Referred to as an **upward** sloping scroll.

522+34

522+33

Primary scroll begins at top and travels downward, this referred to as **downward** sloping scroll.

Top illustration has **stippled** background.

Lower illustration exemplifies heavier **shading** than usually encountered.

Sunburst. Series of lines radiating from center. Usually intermixed with secondary decoration between major rays.

Sunburst on backstrap of a revolver.

Note elliptical major arms, then background shaded lines with some lines terminating in series of stamped hollow circles. **Foliate** pattern at top of backstrap.

572+70

522+70

Sunburst within oval as appears at top and bottom of backstrap.

Spray. Series of arms radiating from a base position. Arms may be straight or curved. Usually with secondary decoration between major arms. Number of arms varies, but always the same number on each side of main vertical arm.

Spray, having major arms of elliptical pattern. Straight thin lines between each arm.

Note **quilted diamond** pattern below spray.

572+75

Sunburst or opposing **sprays** having somewhat floral appearance.

No secondary decoration between arms.

Note lines of small **triangles** that dissect main pattern.

522+41

Rosette. Circle, within which flower petals are usually the dominant feature. May have shading, stippling, or sunburst as secondary features.

Castellated Band. Series of blocks, usually with top rounded corners. May have secondary decoration such as shading.

The above illustration portrays both the **rosette** and **castellatio**n. This is the side engraving on Sharps pepperbox (**115+**) serial number **295**.

Shell. Popular as a decorative motif for long time. Used on guns as early as the queen Anne period.

Below left: Backstrap, immediately behind hammer. Smith & Wesson Model One, 3rd Variation, **9841** (**572+77**).

Below right: Top rear of barrel. Manhattan Model One, 1st Variation. **1427** (**536+2**).

Scimitar. Design that resembles the clipped end on blade, width of blade wider at front. Always seen in series, not independently. Usually shading within each of the "blades." Note **shaded scallops** within illustration

584+14320

513+269

Saw tooth. Wide band, one side of which is composed of a series of triangular cuts.

Shaded C-scroll pattern also seen with illustration.

Fleur-de-lis. Seldom seen on American guns. Pictured is breech of German copy of the Stevens Lord's Model.

372+5030

507 Ethan Allen 507

1727

Only one engraved specimen of the flatbase grip variant has been seen; as illustrated on this and the following page. As the grip contour and corporate name changed, many pieces with conventional engraving are encountered. Refer to Forehand & Wadsworth side hammer revolver (D525) for examples.

The engraving is very shallow and much worn. Patterns do not conform to those popular during the period of manufacture. Whether factory or privately done is unknown, but probably the latter is the case.

The cylinder has a band of maple leaves, each a bit different from the others.

Frame sides and bottom of backstrap have foliate patterns with, in many instances, have webs between segments of each leaf.

A rather nice geometric pattern appears on the cylinder topstrap, this having a large ellipse as centerpoint. Unfortunately, it is well worn and some details are lost in the picture below.

Right and left sides of frame. Note quasi C-scroll in vicinity of screw on left side.

To left is engraving on backstrap, immediately behind shoulder of frame. Note series of stamped small zig-zag lines woven into the pattern.

The geometric pattern approximates a flower with radiating leaves at base.

100 American Arms Company, Wheeler's Patent 100

Serial Numbers 1351 and 669

Engraved examples are frequently encountered. The standard patterns found consist of foliate C-scrolls as seen on serial number 1351, illustrated above. Frames and backstraps decorated, barrel cluster unadorned.

A well executed and unusual combination of patterns may be seen on the following illustration, this being serial number 669. The center of the sides of the frame have the usual foliate C-scroll. A series of pointed leaves border the front and rear areas. Such a motif has not been observed on other handguns of the period, and it is definitely a departure from the commonly used motifs. The backstrap also has a patter not seen elsewhere.

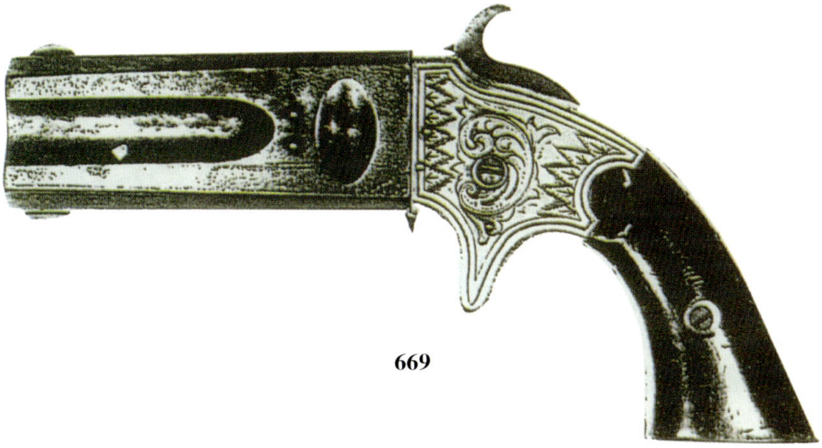

669

30

70

Pattern # + DESCRIPTION		LOCATION	SERIAL #
30	Upward sloping foliate shaded c-scroll; ornate. Main scroll above and to rear of screw. Stippled background. Periphery border line. Topstrap has rearward sloping shaded triangels with bottom border of shaded scallops. Bottomstrap has single wavy line, bordered above and below with series of shaded scallops.		727 10059 3102
31	Upward sloping floral spray. Stippled background. Periphery border line.		14667
32	Upward sloping shaded c-scroll. Main scroll above and to rear of screw. Simple pattern, not as ornate as Pattern 30. Stippled background. Periphery border line.		30950
33	Upward sloping shaded c-scroll. No periphery border line. Main scroll above and to rear of screw. Stippling only within pattern, blank space near edges. No engraving on sides of cylinder top and bottom straps.		317
34	Upward sloping shaded c-scroll, stippled background. Series of small triangles compose periphery border. Half flower with radiating petals on recoil shields. Cylinder straps fully covered with dots, triangles, scrolls.		382
35	Upward sloping shaded c-scroll similar to 10059. Stippled background in upper area, diamond pattern with center dots in lower area. Cylinder topstrap has series of leaves sloping forward.		3102

10059

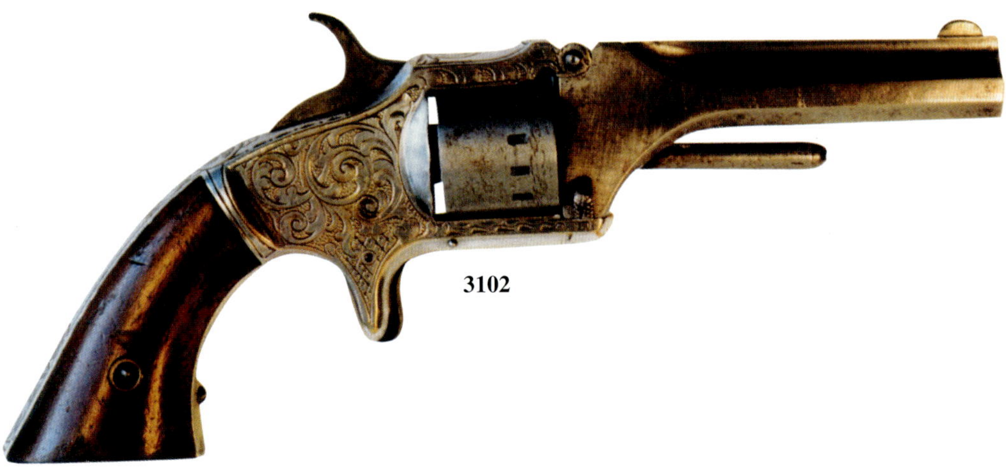

3102

382

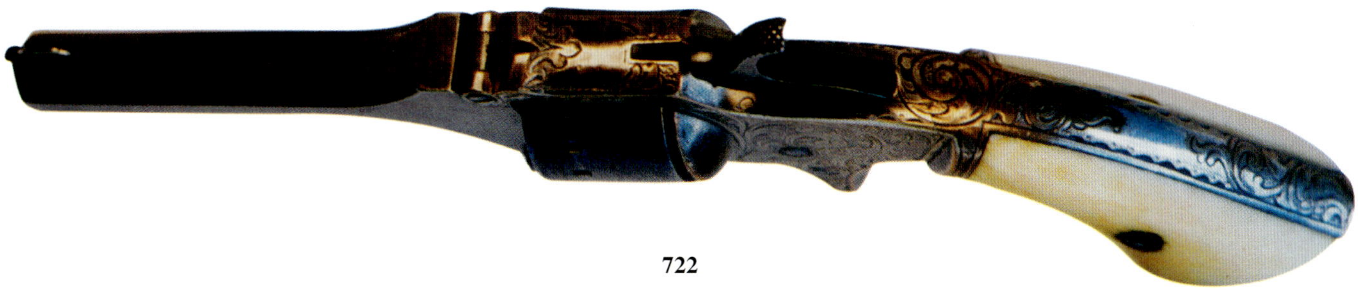

722

Pattern # + DESCRIPTION		LOCATION	SERIAL #
70	Foliate shaded c-scrolls at top and bottom; small unadorned area in center. Stippled background. Series of shaded scallops with hollow circles border edges extending to each side of hammer.		727
71	Foliate lightly shaded c-scrolls at top and bottom, small unadorned area in center; light stippling within pattern. Thin straight periphery border line. NB: pattern continues, fully covers buttstrap.		10059 317
72	Deeply cut sunburst immediately behind hammer. Eight rays; no shading or secondary lines between rays.		
73	Tight foliate shaded c-scrolls from immediately behind hammer to butt; small unadorned are in center. Periphery border composed of shaded scallops surmounted by a thin wavy line.		382

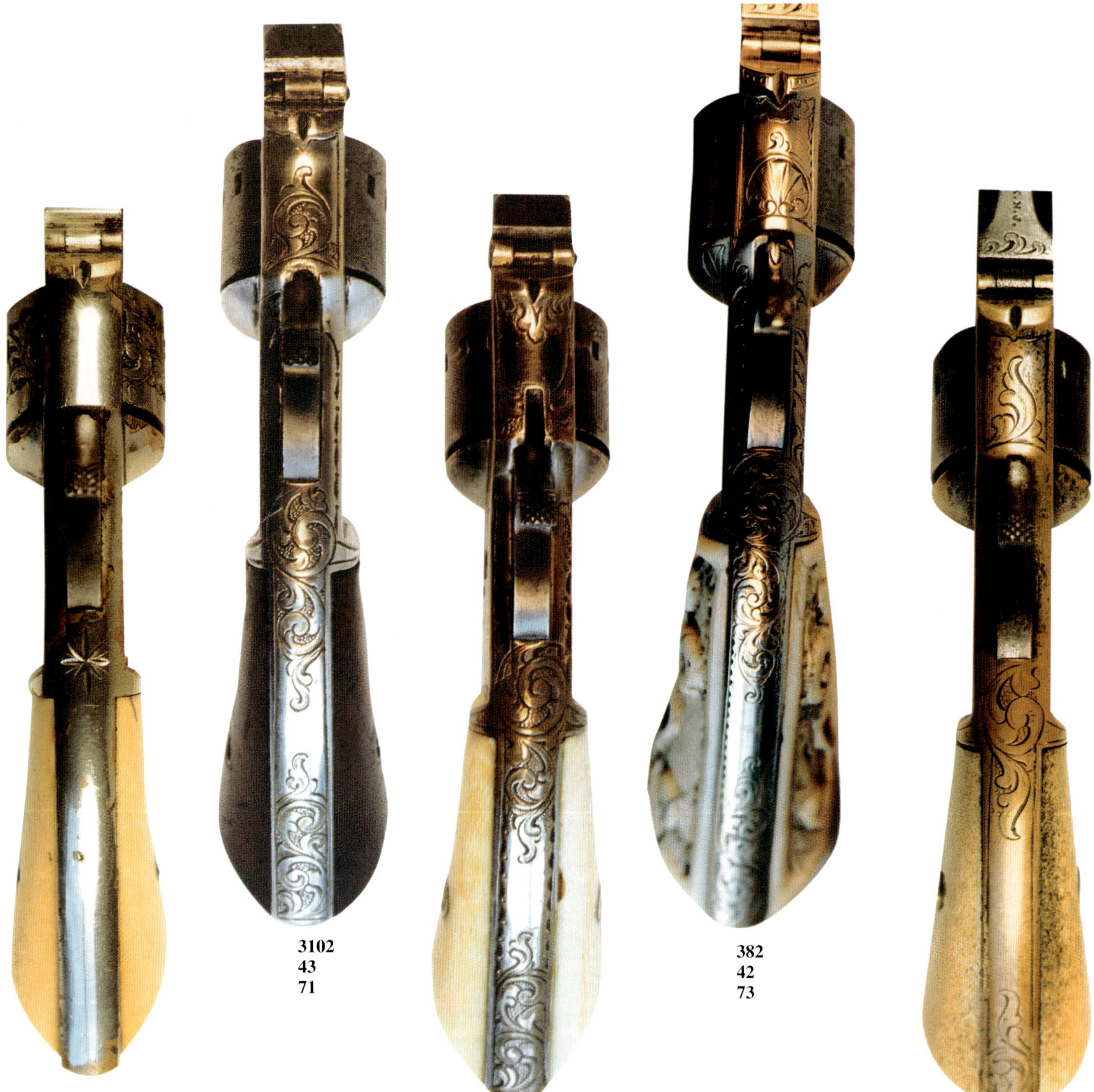

3102
43
71

382
42
73

317
0
72

727
40
70

10059
41
71

382

3102

727

10059

317

Serial Number **382**. This is the only example observed with engraved hammer. Patterns on both sides, also on top front of hammer.

Note also engraing on rear of recoil shield; another aspect seldom seen.

Buttstrap engraving on **3102**

Buttstrap engraving on **382**

727. Initials J W A stamped in center. In longhand inside left grip: J.M. Arnold Bought. A.R. Aronold was the Superintendent of the Manhattan Fire Arms Company; J.W. may have been a relative.

382

727

3102

10059

Pattern # + DESCRIPTION		LOCATION	SERIAL #
40	Shaded petals with stippled background.		727
41	Slightly shaded leaves, no stippling. Periphery border line follows contour of topstrap.		10059
42	Shell at rear, few shaded leaves at front. Dots within straight line borders each side of sight notch. Shaded scallops surmounted by thin wavy lines on sides.		382
43	Foliate c-scroll, shaded, stippled background at rear. Periphery border line follows contour of topstrap.		3102

513 Bacon Manufacturing Company 513

269

All revolvers of this model were engraved with the patterns pictured herein; there were none without this form of engraving. The pattern was also used on larger calibre models, and earlier on some of his percussion pepperboxes. The pattern was not used on his contemporary cartridge pepperbox (none of which have been observed with any type of engraving). The firm was in operation from 1858 to 1866.

Thomas Bacon left the firm in 1864 and established the Bacon Arms Company. The products made by the new firm were not of the quality of the Manufacturing Company and, when found engraved, were of a different pattern and not well executed.

Only the sides of frames were engraved, and then with a quite uniform C-scroll pattern. Small semi-circular cuts appear within the motif as shading. No engraving on areas other than sides of frame have been noted.

269

Note the periphery border line which appears on sides and top of frame. This is a distinctive feature of the Bacon Manufacturing Company revolvers. Also found on some of the earlier percussion pepperboxes made by Bacon before the Manufacturing Company was formed.

The saw-toothed line at rear forms a V at top center of backstrap.

366 Biederman, Multi-Barrel 366

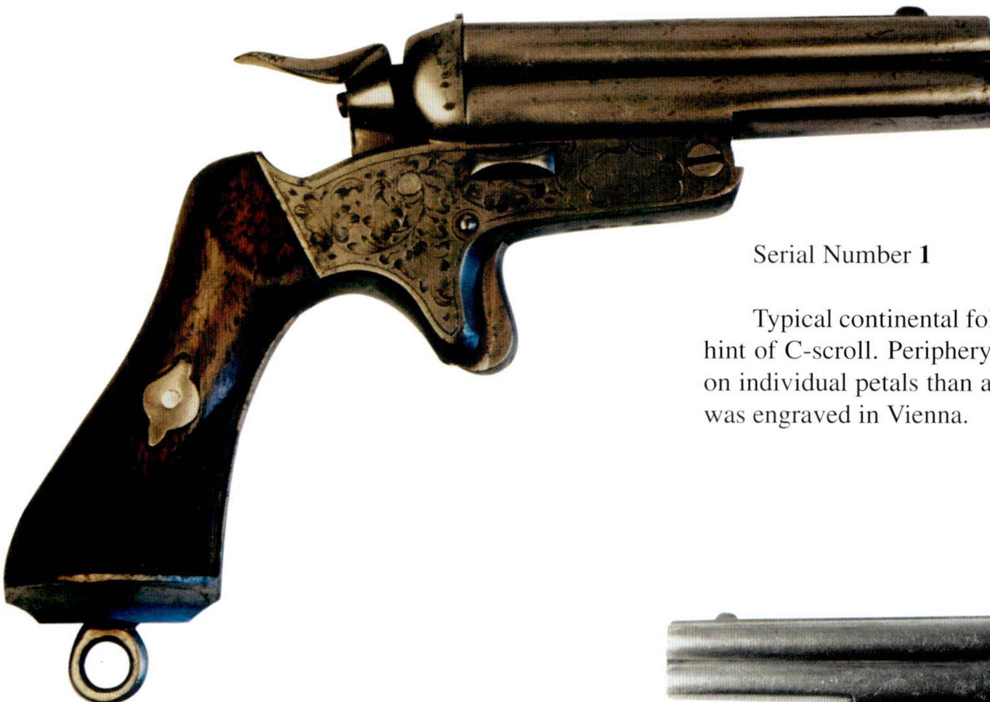

Serial Number **1**

Typical continental foliate pattern, this example having only a hint of C-scroll. Periphery border lines and light shading. Accent on individual petals than a tight group of folliation. This example was engraved in Vienna.

300 Biederman, Continental Copy of Smith & Wesson 300

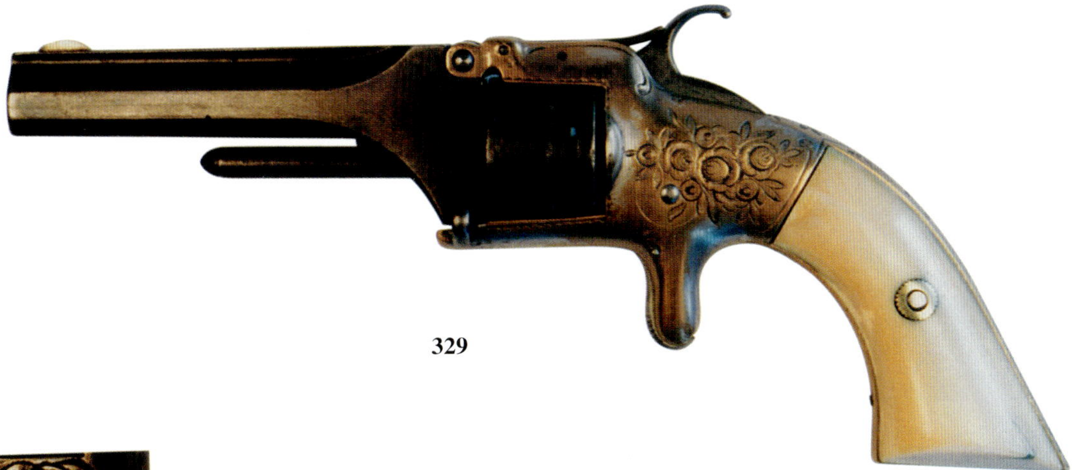

329

A rather close copy of the Smith & Wesson First Model First Variation revolver. Brass frame, oval contour through action area. Unlike the original, the hammer is of one-piece construction.

Barrel rib is marked: SMITH & WESSON, SPRINGFIELD, MASS. Other than serial number and crown over U and V proofs, no other markings. Proof marks similar to some German of the period; not remotely close to those of Belgium, France, or England.

Flower and leaf engraving on frame sides, flowers and geometric patterns on backstrap and buttstrap.

Buttstrap.

306 Biederman, European Copy of Smith & Wesson 306

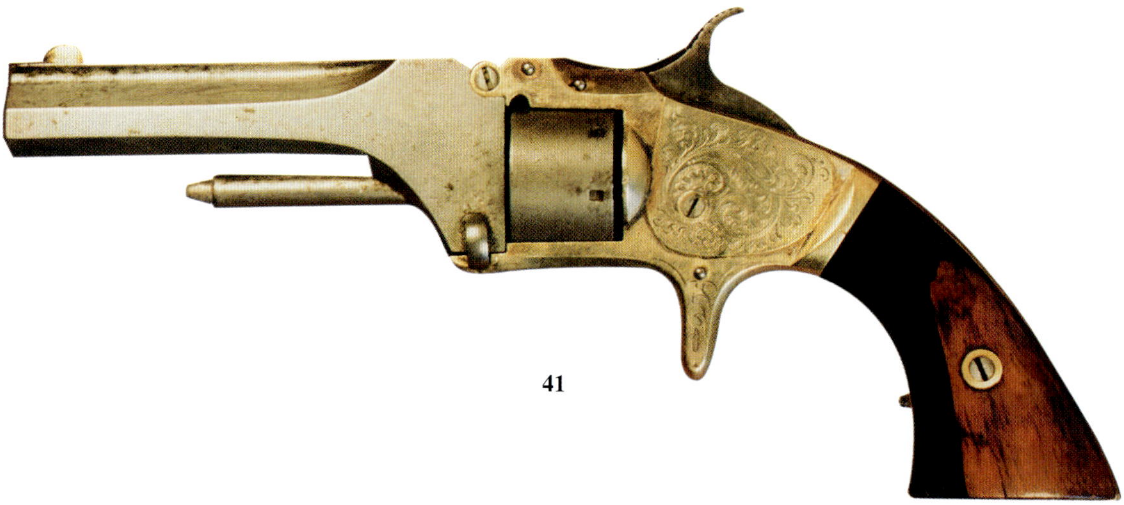

41

Very similar to the Smith & Wesson Model One Second Issue; was probably made in limited quantity in the late 1860s. Unique feature is a solid German silver frame.

Other than serial number, no inscriptions or proof marks. Believe of European manufacture for following reasons: type of engraving; nearly identical to other copies that bear European proof marks; two-tone rosewood stocks as found on mainly European pistols.

Engraving is of the common foliate C-scroll as found on both European and American firearms of the period.

Backstrap pictured to left.

Note shell designs within some of the scrolls.

Scrolls and folliation devoid of shading or stippled background, features found on most engraved American firearms.

304 Biederman, British Copy of Smith & Wesson 304

2

A close copy both mechanically and in outline of the Smith & Wesson Model One Second Issue revolver. No indication of manufacturer; bears British proof marks. Probably made in the 1860s.

Webley claimed to have made similar pieces with permission of Smith & Wesson, but no record of Smith & Wesson having done so has been located.

Tight well executed foliate C-scroll pattern on sides of frame. Backstrap, buttcap and barrel also engraved. Three lines encircle front of cylinder near edge. Central line is wider, has opposing Vs cut into both sides, this pattern resembles a spiralled ribbon. It is also found within the periphery border line near bottom of sides of frame, on top of frame parallel to hammer on both sides, encircles the buttcap, and encircles sides of barrel near muzzle.

A series of shaded petals is engraved into both sides of cylinder bottomstrap, below line of opposing Vs.

304 Biederman, British Copy of Smith & Wesson 304

321 Biederman, Belgian Copy of Smith & Wesson 321

3

Belgium (Liege) proof marks and serial (assembly?) number in several places; no other markings. Very well made.

Other than contour of ejector rod and smooth barrel release wings, this is a very close copy of the Smith & Wesson First Model Third Variation revolver. At one time, all metal was blued; frame is now without finish.

Gold periphery border lines on sides of frame, surrounding flutes on cylinder, on top of barrel rib, and on full length of backstrap.

Quilted diamonds a ellipse between each flute on cylinder, these in gold.

Gold inlaid diamond at rear of barrel rib, bordered by silver inlay at back.

Sides of frame have very open foliate C-scrolls in gold, small solid circle in gold at top rear.

Alternating gold and silver diamonds and ellipses, innertwined, on top of backstrap and on buttstrap.

515 Colt's PT.F.A. Manufacturing Company 515

Illustrations on this page are of serial number 24480. This is the earliest serial number personally observed, the number sequence being carried forward from earlier model with integral ejector rod housing (D514). It has the early high spur hammer as opposed to the much more common low spur hammer of later standard production. The Open Top, or "Old Model", including ejector rod variant, was produced from about 1871 to 1885.

Engraving probably done after the revolver left the factory as greatly varies from the normal Colt motifs. The open petals of flower with leaves radiating from background is not common on Colt revolvers. The design is similar to that found on the Reid revolvers with Holey Manufacturing Company inscription; refer to page 123.

All engraving is rather crudely done. Conventional foliate patterns are found on sides of barrel and near top of backstrap.

A band of joined rectangles encircles the rear portion of cylinder, front and rear edges having triangular slots cut into each.

90362

Revolver illustrated on this and next page has the typical patterns found on engraved examples of this model. It is of the common low hammer variation.

This model is most often found with full nickel plating. Grips may be walnut, ivory, or pearl. Quality of engraving is mediocre.

Foliate clusters with patches of stippling are found on the rear portion of cylinder. A band of double line connecting diamonds with crosses in center of each diamond will also be found on cylinders of this model.

A wavy line with hollow punched circles between each wave borders each side of the inscription. The length of inscription between the Maltese crosses will vary; the address line may be of either upright or slant block capital letters.

Sides of barrel have series of rectangles, each with a hollow punched circle in center. Top and bottom edges have double lines of shaded scallops. A shaded spray appears forward of the blocks. This pattern on side of barrel is used quite consistently.

Backstrap has radiating spray at top, followed downward with double-line triangle, ellipse, diamond, ellipse, smaller ellipse. Solid punched dots within and bordering lines.

No engraved examples of the Colt Open Top with integral ejector rod housing on barrel have been seen or reporting, though may exist.

Refer to page 179 of the Locke Collection book for illustrations of typical engraving found on the Colt Open Top revolvers.

517+ Colt New Line 517+

Illustrations on this page are of serial number **21077**. This revolver is housed in original wooden case. Ivory grips; few parts gold washed.

The usual patterns seen on this model employ the foliate C-scroll, follow conventional patterns used by other gun makers. The cylinder and barrel decoration varies. Cylinders commonly have punched geometric patterns or foliate clusters between flutes, not very well executed.

Nickel plating on engraved pieces is standard. Some will be found with gold wash on some parts. When engraved, grips usually of pearl or ivory. Refer to Locke Collection book, pp. 188 for illustrations of several engraved examples.

520 O.S. Cummings 520

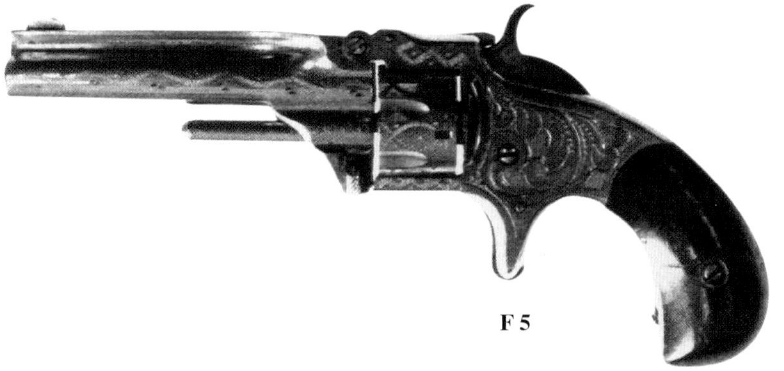

F 5

O.S. Cummings of Lowell, Massachusetts, produced a revolver quite similar to the Smith & Wesson Model One Third Issue.

Production of this iron frame copy was quite limited, as was the quality of the gun and the engraving. When marked, the following inscription appeared on the barrel rib: O.S. CUMMINGS, LOWELL, MAS (note only one S in MAS).

Engraved examples utilized the foliate C-scroll pattern on sides of frame. Some had crude zig-zag punching on cylinder, cylinder top and bottom straps, and on barrel. Punched dots were applied in various places.

Revolvers with CUMMINGS & WHEELER, LOWELL, MASS. inscriptions exist; they vary slightly from the Cummings, share the same poor quality in all respects. None have been observed with engraving. Refer to D520 and D521 in "Notebook" for background information and relationships with others in the firearms business.

522+ Deringer 522+

6094

The Deringer Rifle & Pistol Works existed in Philadelphia, Pennsylvania from the early 1870s to 1880. Dr. Jonathan Clark, Henry Deringer's son-in-law, and Charles Foehl, who patented the basic pattern, were the principle figures in the firm.

Three basic models, with a total of five variations, were produced; breakdown appears below. This was a small company with very limited production compared to some of the other manufacturers of the time.

Probably local engravers were used. Many of the patterns used were of the standard type, mainly foliate C-scrolls. However, some fine quality engraving with motifs varying from the norm have been encountered.

Variances in "Notebook", with corresponding Description Numbers: Model One, Type One = D522; Model One, Type Two = D522.1; Model Two, Type One = D523; Model Two, Type Two = D523.1; Model Three = D524.

No engraved examples of Model One, Type Three, or Model Two, Type One have been seen, though probably exist. When engraved, all major areas are decorated; barrel, cylinder, topstrap, and backstrap. Standard finish is nickel plating. Some with silver and gold washed cylinder, hammer and trigger exist.

6094

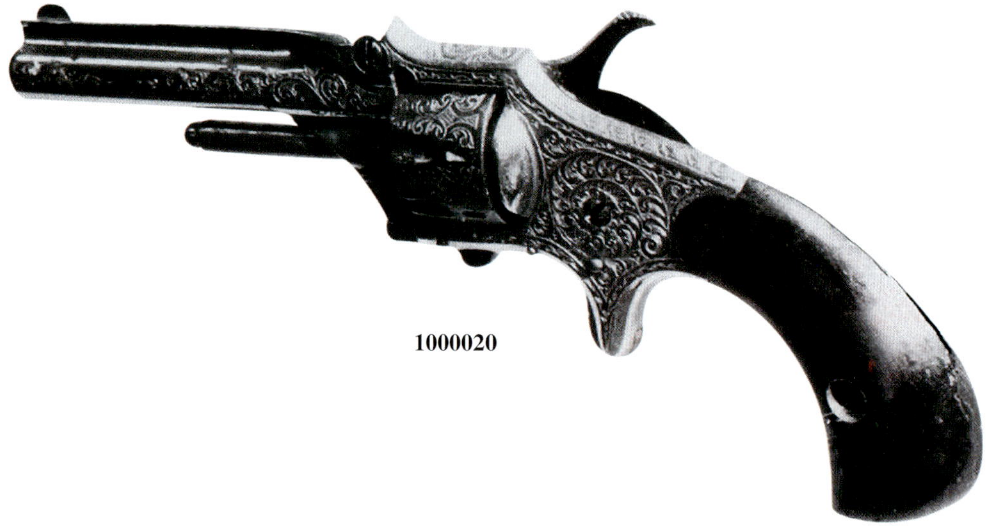

1000020

During the nation's Centennial Celebration of 1876 held in Philadelphia, some Deringer revolvers were engraved, presumably for the event. The few such pieces encountered were very finely engraved.

Apparently a separate serial number range was used (pictured below) completely unnumbered, and that in the upper right hand corner has what seems to be a rather impossible number. Possibly the range began at one million?

Pattern # + DESCRIPTION		MODEL	SERIAL #
1	Series of blocks, circle within each block. Blocks separated by three vertical parallel lines. Shaded scallops above and below blocks, form border. Rear area with diagonal band - double straight lines at edges, double wavy line in center. Foliate sprays each side of band, stippled and shaded.	1-1	25
2	Series of circles on flat. Foliate C-scroll, stippled background, in rear area.	1-2	478
3	Barrel flat is not decorated. Foliate C-scroll, stippled background, in rear area.	1-2	708
4	Quilted diamonds in panel at front, then quilted diamond then hollow diamond. Small sprays protrude from junctures. Shaded floral area at rear, includes quilted diamonds. Rear portion through which ejector rod is housed has series of chevrons. Bottom of rear area has foliate C-scroll.	2-2	(none)
5	Sunbursts within ovals, separated by foliate sprays. Some stippling within pattern. Series of small circles within rib flute.	3	6094

Model Two, Type Two (no serial number) Full view on preceding page.

Pattern # + DESCRIPTION		MODEL	SERIAL #
10	Border line surrounds inscription on top of rib; small loop at front, widens at rear. Rear line composed of shaded scallops. Cylinder topstrap has opposing sprays to right, left, front and rear. Double straight lines cross diagonally, form an X through base of sprays in center of topstrap.	1-1	25
11	Top of rib plain except impressed inscription. Cylinder topstrap has opposing sprays to right, left, front and rear; small hollow circle at base of each spray. Curved dissecting lines composed of small triangles form an X which dissects sprays.	1-2	478
12	Top of barrel rib has impressed inscription only. Cylinder topstrap has shaded leaves; stippled background.	1-2	708
13	CENTENNIAL engraved in slant block capitals on rib. Quilted diamonds in arrow near muzzle. Foliate C-scroll pattern at rear, surmounted by spray of leaves; quilted background. Single periphery line bordering top of rib.	2-2	none
14	Three ellipses and diamond at front. Series of shaded scallops border the impressed DERINGER PHILADA inscription. Spray at rear with secondary thin lines between arms.	3	6094

478
31
61

Pattern # + DESCRIPTION	MODEL	SERIAL #
30 Upward sloping shaded foliate C-scrolls. Main scroll above and to rear of screw. Stippled background. Periphery border line. Series of solid scallops and small circles on top and bottom straps.	1-1	25
31 Foliate shaded C-scrolls radiating from rear screw. Stippled background. Periphery line. Wavy lines with small circles between waves on top and bottom straps.	1-2	478
32 Upward sloping foliate C-scrolls with crosshatching and shading. Main scroll directly above rear screw. Stippled background. Periphery border line. Series of scallops and small circles on top and bottom straps.	1-2	708
33 Upward sloping foliate C-scrolls. Main scroll above rear screw, with offshoots. Parallel line background; no stippling. Leaf periphery lines which extend to and cover top and bottom straps.	2-2	none
34 Upward sloping foliate C-scrolls. Main scroll above and forward of screw. Slight shading, stippled background. Periphery line. Series of shaded petals on top and bottom straps.	3	6094

25
30
60

25
10

478
11

708
12

none
13

6094
14

Pattern # + DESCRIPTION		MODEL	SERIAL #
40	Sunburst superimposed on large double line.	1-1	25
41	Large X composed of double line of small triangles, slight curvature of arms. Five-branch sprays between arms of X, small circle at base of each spray.	1-2	478
42	Shaded leaves extend forward; some border stippling.	1-2	708
43	Tight foliate C-scrolls.	2-2	none
44	Foliate C-scrolls, stippled background. Curved single line at front; sides and rear have series of shaded scallops; small triangles between scallop tops.	3	6094

25

Pattern # + DESCRIPTION	MODEL	SERIAL #
70 Sunburst within oval surmounted by petals behind hammer. Small triangles over shaded scallops border hammer. Sunburst within oval, then spray at base of backstrap. Single border line on each side of backstrap.	1-1	25
71 Double and single ellipse, small circles border. Line extends forward on sides of frame by hammer, terminate in two teardrops.	1-2	478
72 Shaded leaves behind hammer; few stipple marks.	1-2	708
73 Tight foliate C-scrolls. Forward slanting blocks each side of hammer, each with shading and notch. Plain bars separate blocks.	2-2	none
74 Diamond with crosshatched center behind hammer, petals above and below diamond. Series of shaded scallops each side of backstrap. Lower area has double ellipse - diamond - two single ellipses. Small circle outside junctures of segments. Series of shaded scallops border frame on each side of hammer; small triangles above scallops.	3	6094

Model Two, Type Two (D523.1)

Pattern # + DESCRIPTION		MODEL	SERIAL #
60	Blocks, each holding two circles, encircle rear area. Four parallel lines between each block. Rear of band has solid triangle at rear of each block. Front of band composed of solid scallops. Band of small triangles encircle front of cylinder. (cylinder has smooth surface, no fluting)	1-1	25
61	Series of solid thin ellipses encircles rear area. Sprays emanate from large ellipses, curves below sprays. (cylinder has smooth surface, no fluting)	1-2	478
62	Foliate C-scroll pattern between flutes; quilted diamonds 2-2 at bottom of pattern; series of Vs within top of pattern. (cylinder has semi-fluted surfaces)	none	
63	Shaded foliate C-scroll between flutes, with stippling. Series of small solid triangles encircle rear. (cylinder has semi-fluted surface)	3	6094

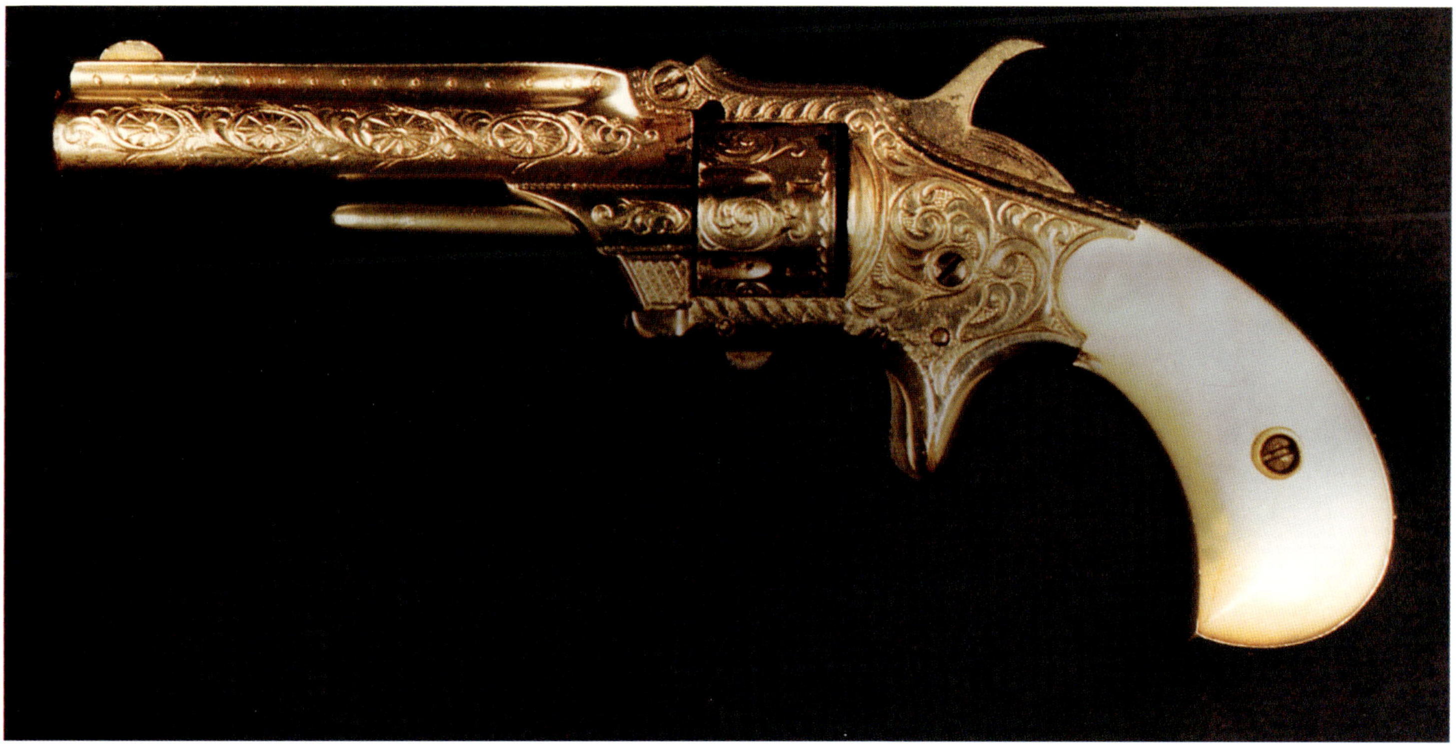

6094

708
42
72

478
41
71

25
40
70

6094
44
74

NONE
43
73

525 Forehand & Wadsworth, Side Hammer 525

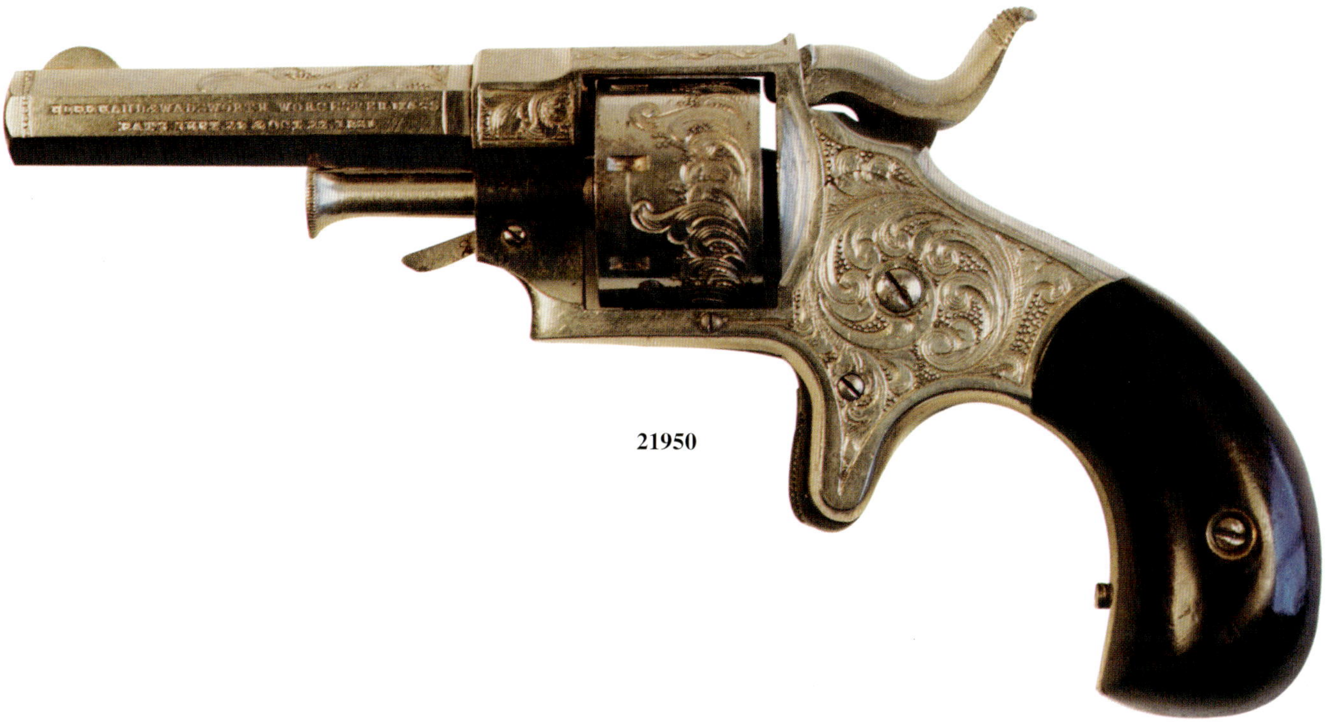

21950

This revolver is the continuation of the basic model produced by Ethan Allen under various corporate names for several years. After the death of Allen the firm operated under the name of Forehand & Wadsworth from 1871 to 1890. Refer to "Notebook" (D525) for discussion of mechanical aspects and corporate history.

Several examples of engraved pieces have been seen, nearly all with nickel plating. A few pieces have been seen with silver barrels and frames with gold washed cylinders; these were not engraved. Grips were of rosewood, ivory, or mother-of-pearl (the latter may have high relief carving on right panel).

22860

22860

21950

No example has been observed with engraving on hammer. Sides of frame carry conventional foliate C-scroll engraving. Patterns on cylinders and barrels vary. Backstraps have a combination of C-scroll and geometric line engraving.

TOPSTRAPS

22860 Series of clipped corner blocks with hollow punched circle in center of each on top flat, these bordered by deeply cut shaded Vs. Forward area has double-lined ellipse bordered by punched hollow circles. Punched wavy line on each diagonal flat.

21950 Extended foliate pattern with clusters of punched berries on top flat. Each diagonal flat has a wavy line resembling a vine with protruding leaves.

BARRELS

21950 has foliate pattern on upper diagonal flats and right side; double line filled with punched hollow circles near muzzle on side and upper diagonal flats, geometric pattern on top.

22860 Delicate floral spray on top; curved double line near muzzle on top, upper diagonal, and side flats. Geometric pattern on right side flat. Upper diagonal flats have double wavy line bordered by shaded scallops.

BACKSTRAPS

Quite similar. Both have foliate C-scrolls at top, then periphery lines ending in double-lined ellipse-diamond-ellipse.

21950 is more ornate, having stippling as background of scrolls, and series of shaded Vs in two places on border lines. Border lines do not extend forward of scrolls on **22866**.

22866 has punched hollow dots bordering top and bottom patterns; **21950** does not.

22860

21950

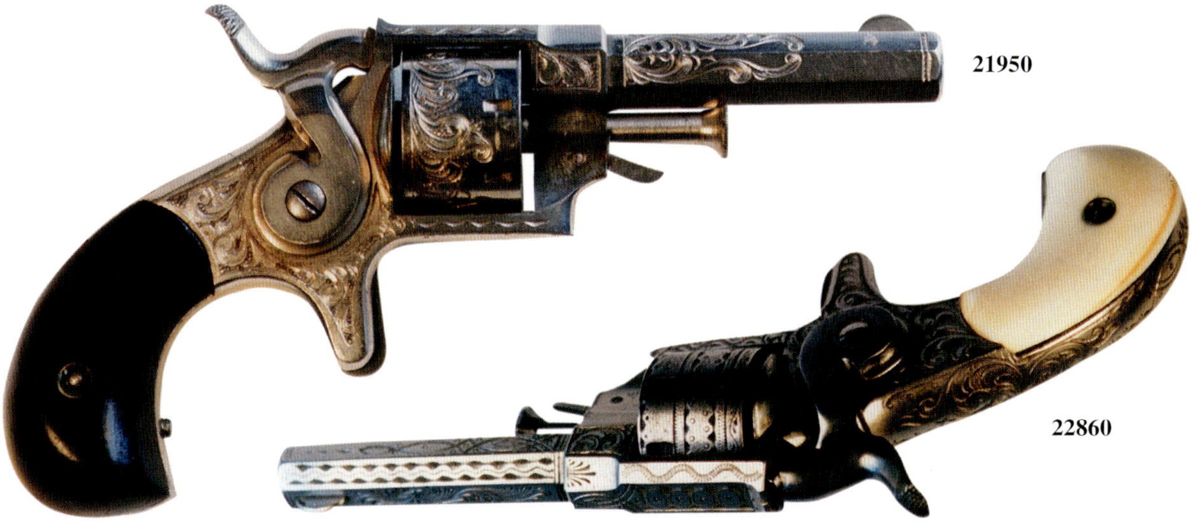

21950

22860

21950

Note the more complex pinwheel arrangement of the foliate scrolls on left side of frame on **22860** as shown at lower right.

22860

357 Sistema Merolla Giovanni 357

Produced in Italy during the latter half of the nineteenth century. Included herein for both an example of Italian engraving and the uniqueness of the piece. No serial number. Unlike other entries in this work, the pistol was never in personal collection; rather, the property of Robert Palmer.

Full description will be found in "Notebook", Description Number 357.

Foliate C-scroll patterns, no stippling in background. Note decorative cuts on heads of screws.

372 Hafner 372

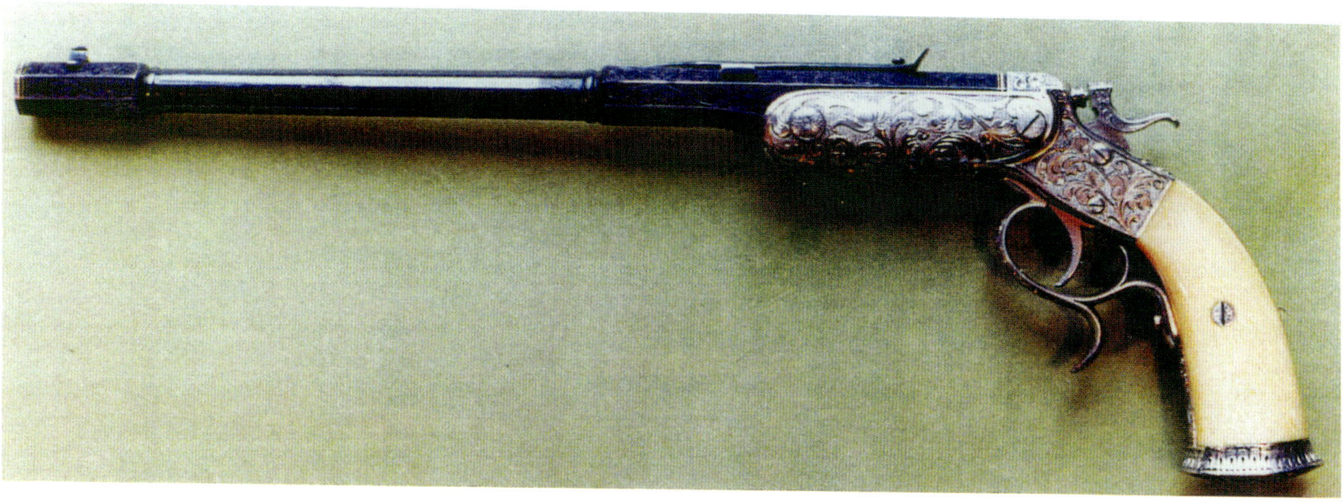

5030

Made by V. Hafner of Cassel, Westphalia, Germany. Cased. The Stevens Lord's Model (D240) was in production from 1880 to 1911. A few pistols of the same design made by German gunsmiths are known. All are of presentation grade. One, with solid gold frame, was presented by Kaiser Wilhelm II in 1894. All of .22 calibre, as that illustrated. They were probably all made near the turn of the century. All are cased, with accessories.

Stevens also created engraved presentation grade Lord's Model pistols. Refer to Kenneth Cope's Stevens book, Plates 26 and 27. Also, German version illustrated in Plate 18.

Engraved monogram on butt plate. Mention is made of other German pistols of this design with the same treatment of buttplate. Foliate engraving is pleasing and quite well done, as is all engraving throughout the pistol.

Nickel plated brass frame, blued barrel with inlaid gold bands. Maker inscription of V. Hafner, Cassel, inlaid in gold on top of barrel.

224 Lombard 224

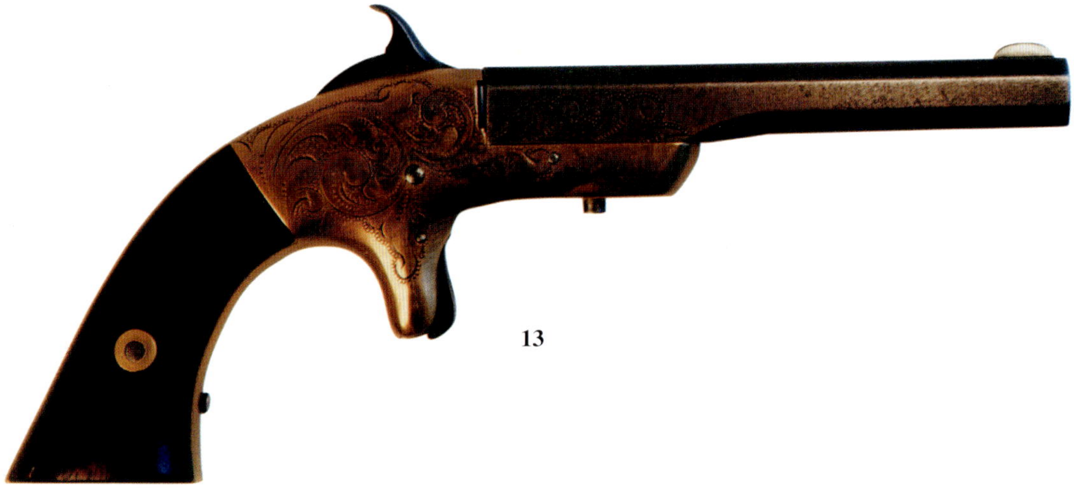

13

H.C. Lombard and Company operated in Springfield, Massachusetts from 1859 to c. 1862. Many of their pistols were unnamed, mostly all were not engraved. They may have produced similar pistols for wholesale and retail dealers using other names (when marked) on barrels.

The foliate C-scroll pattern with stippled background as common during this period was used. Note in picture below the extended leaves flowing forward from side of frame to rear of barrel.

Backstrap.

Buttstrap.

536+ Manhattan F.A. MFG. CO., Model One 536+

This first cartridge revolver of the Manhattan Fire Arms Manufacturing Company was produced from 1860 to June of 1861 when the Second Model (D539, page 37) was introduced. Slightly over 8200 were made.

Barrel engraving was quite common, probably 40% of production was so treated. Only rarely will engraving on sides of frame be seen. Two patterns of engraving were used at back of barrel rib. Some examples have no decoration or company name on barrel rib.

An unusual feature, found only on the First Model Manhattans, was the seemingly randomly placed copper inserts in the frame. These may have been used to cover casting flaws. Several examples have oval copper plates, sometimes engraved, set into the top of backstrap. These were definitely for decorative purposes, not simply to cover flaws.

No examples of engraving on cylinder have been observed. The engraving on sides of barrels is of a quite consistent pattern.

Only a few of the earliest production had brass frames, the great majority had iron frames. Finishes of full blue or full silver were used. No nickel plated First Models have been seen.

Three variations of the Model One exist; they are described in "Notebook" under Description Numbers 536, 537, and 538.

1427 First Variation

1427
5

6352
5

1427
2

6352
1

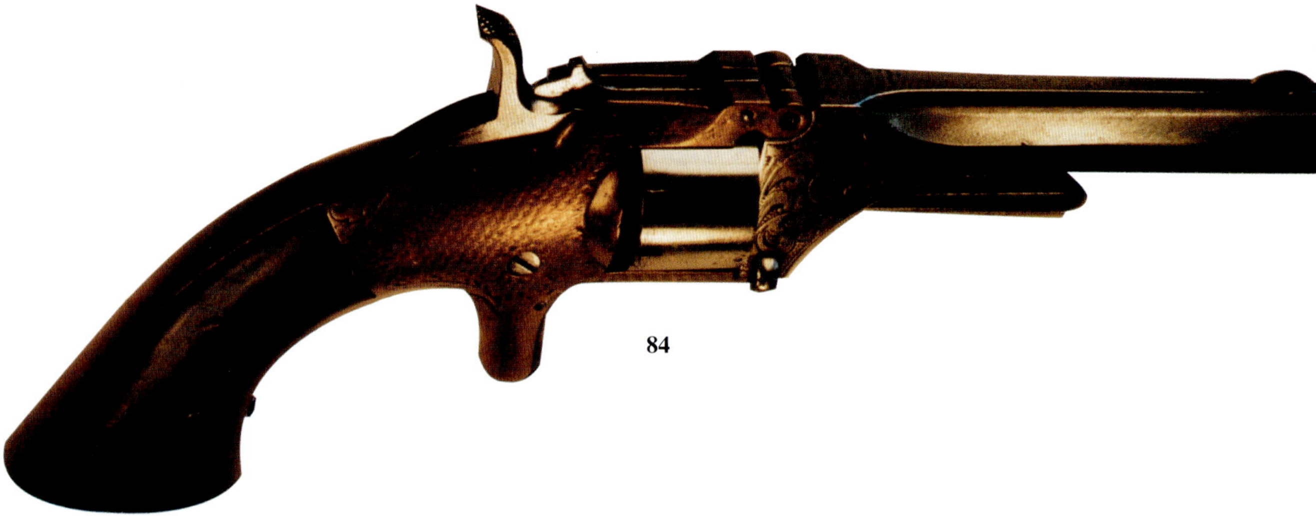

84

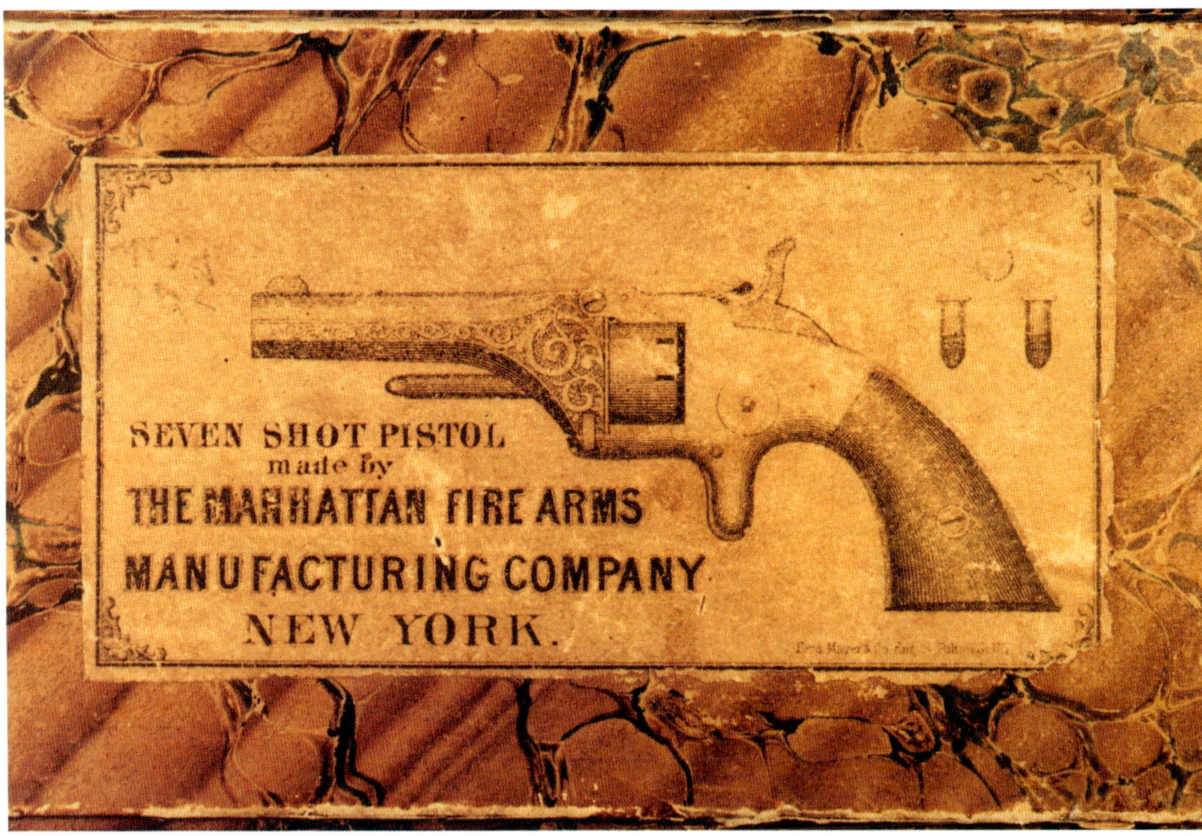

Note engraving on barrel of revolver and label. Cardboard box, partially shown above, with marbleized overlay, came with serial number **1427** which is pictured below.

539+ Manhattan F.A. MFG. CO., Model Two 539+

Serial Number 4364

Engraving coverage on this model varies:
(1) Most common style was engraved barrel, frame plain;
(2) Sides of frame, backstrap and buttstrap engraved, barrel plain;
(3) Sides of barrel and frame engraved, plain backstrap.

Names or initials of owners may appear on buttstrap. On those examined so marked, it is believed the inscription was done by same person at time of engraving of the entire piece.

Silvered frame with blued barrel and cylinder. After corporate name was changed to American Standard Tool Company (D508), finish was full nickel plating.

A pattern unique to Manhattan, found on both Models One and Two, is a basket background on sides of frame. It is believed the "basket" pattern was pressed into the areas to be engraved, then scrolls were hand cut within the basket pattern. Under high magnification, the "weave" of the background is so uniform that hand engraving of pattern is very improbable. Four examples have been closely compared, the "weave" is exactly the same on all, but hand engraving of scrolls within the weave varies from piece to piece. The Model One with basket background is of a different pattern, as is the engraving.

No engraving is found on top of barrel or on cylinder topstrap.

A roller engraved series of vignettes is found on cylinder which consists of:

man with pistol shooting Indian holding rifle;
man with pistol shooting Indian holding tomahawk;
woman with pistol shooting Indian
woman shooting Indian about to scalp another woman.

There is also a continuous floral pattern encircling front of cylinder.

Pattern # + DESCRIPTION		SERIAL #
30	Shaded C-scrolls with basket weave background. Scrolls and weave extend over surface of topstrap and bottomstrap.	4196 4289 4350 4364
31	Upward sloping shaded foliate C-scrolls. Stippled background. Main scroll above and to rear of screw. Periphery border. Topstrap has series of rearward sloping triangles with line of shaded scallops at base. Bottomstrap has single wavy line with shaded scallops above and below. Swell at rear to meet contour of grips has small petals within single line periphery border.	3048

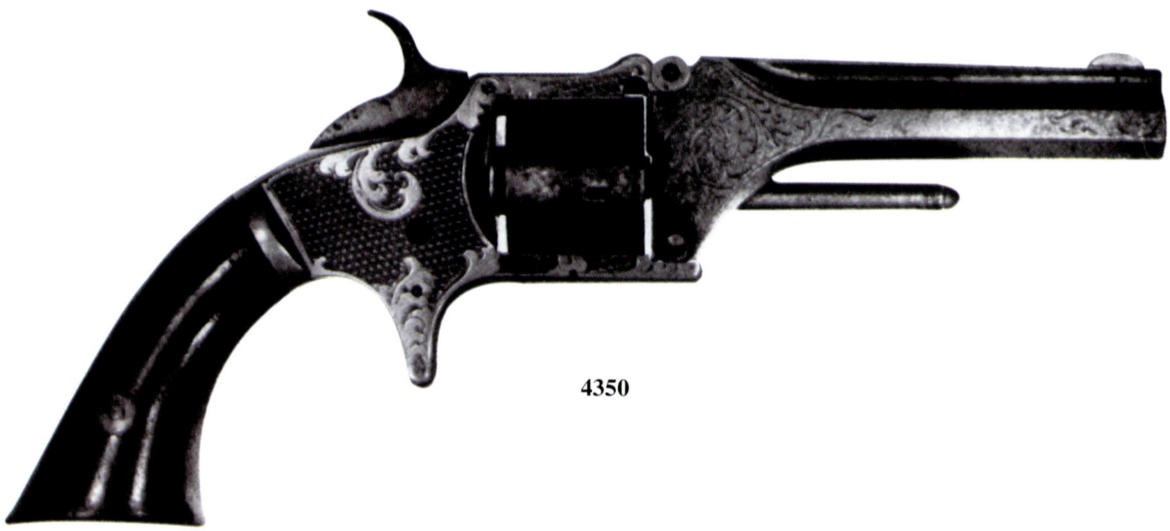

4350

Pattern # + DESCRIPTION, BARREL ENGRAVING		LOCATION	SERIAL #
1	Foliate shaded C-scroll, stippled background Plain, no decoration. Small foliate shaded C-scroll with stippled background.	side, rear top side, front	3048 4196 4289 4350 4364 5851

4196

4289

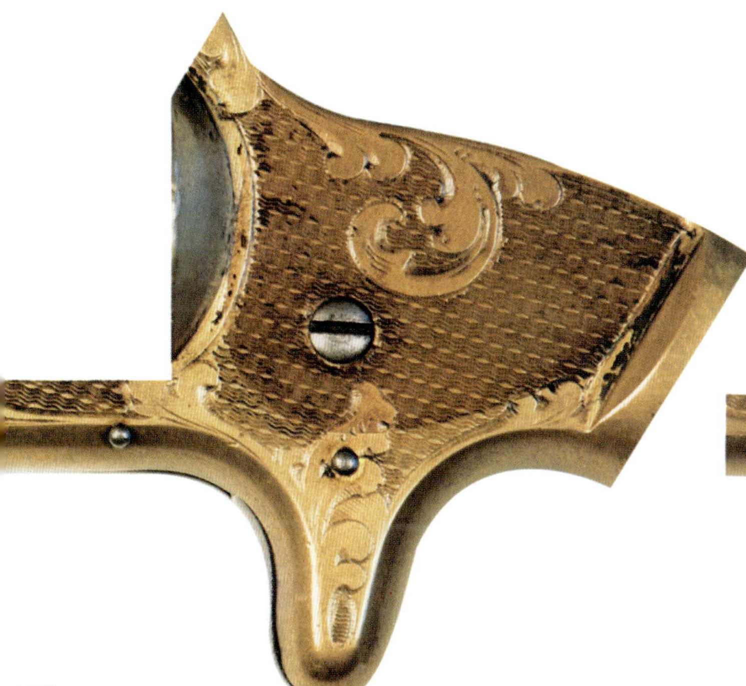
4350

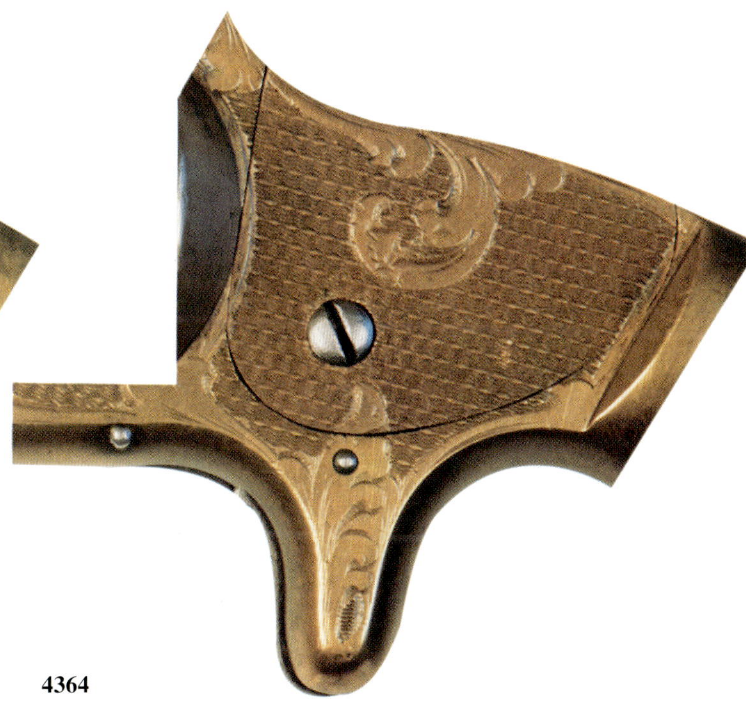
4364

Closeup of 4350. Note uniformity of stamped background pattern. This basket weave was also used on a few buttstraps of the Manhattan Model One, and on one frame of that model. Note engraving cuts on scrolls and rear edge.

225 Marlin, Model One 225

Serial Number **171** is pictured above and below. Top barrel bears the Marlin inscription, an element seldom seen on engraved examples.

There seems to be no uniform style of decoration on the few engraved pieces observed. As is the case in general, engraving was probably executed for large sporting goods firms rather than by the factory.

Some engraved pieces will have decoration on top barrel flat where company inscription would normally be placed. Whether these pieces were ever named cannot be determined. Most engraved guns, without engraving on top barrel flat, are unnamed.

Those recorded as having engraving do not fall into any sequential pattern, nor do they appear within a set block of serial numbers.

Stamped zig-zag lines are usually present on frame sides near standing breech and frame extension. This may also be found on side flats of barrel. In general, the engraving on this model is rather mediocre.

Note the stamped zig-zag lines (below) that curve inward and form a diamond at juncture; a feature found on barrel vertical flats and on backstrap.

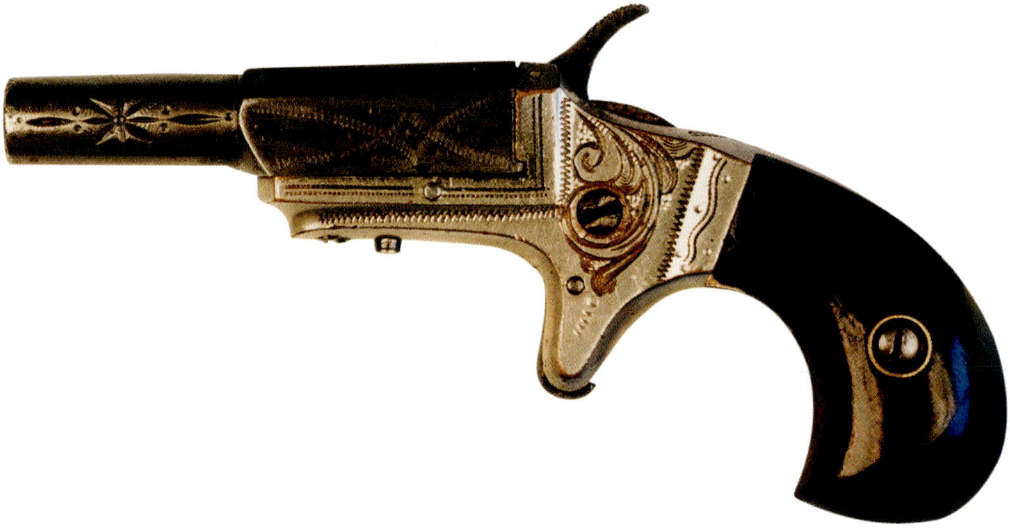

2901

171

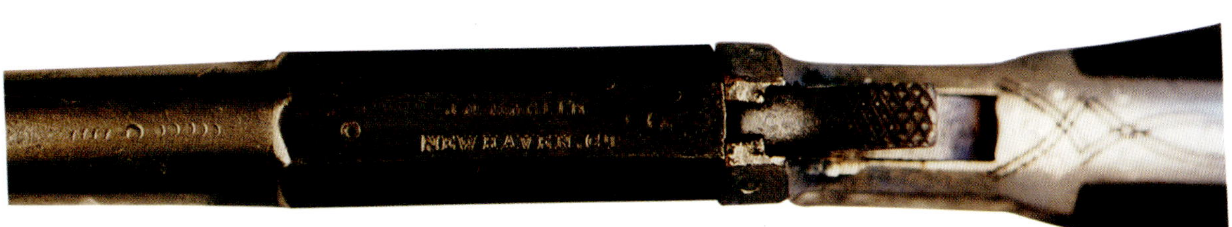

540+ Marlin Revolvers 540+

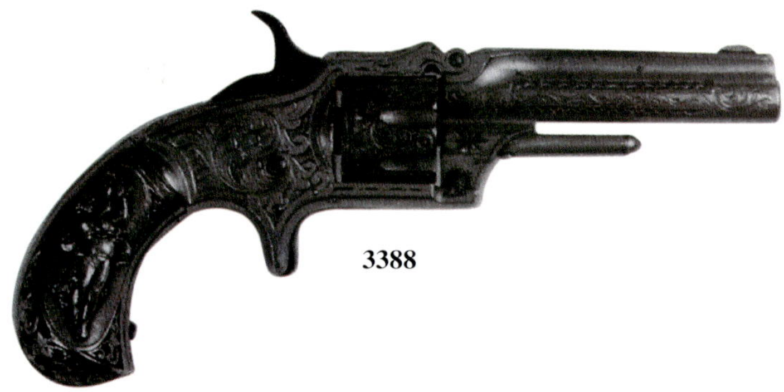

3388

Two basic models were produced in .22 calibre. Both will be found engraved. The engraving is typical of the period with conventional C-scroll patterns; barrels, backstraps and cylinders have different patterns also common during the period. The quality is mediocre in most cases. Grips may be of rosewood, hard rubber, ivory, pearl, and rarely the electro-type DeGress creations.

The model with Smith & Wesson type tip-up barrel, similar to the American Standard Tool Company revolver, was made variations under the label of XX Standard (Description Numbers D540 and D541 in "Notebook"). The first type with ribbed octagonal barrel was produced in small quantity in 1873 and 1874. The second, most common, model as shown to right, was made from 1875 to 1887.

4650

Two engraved examples of the LITTE JOKER are pictured above.

This example is serial number **104**, is fully nickel plated and engraved.

Stocks were made for Marlin by F. DeGress and are quite rare; design on right side is a full length woman loosely wrapped in a length of cloth. All decoration in both grip panels is of higher relief.

The second model revolver had a solid frame, was probably created to compete with the cheap suicide special type revolvers which began to appear in the 1870s. One variant was labeled "O.K." (D542 in "Notebook"), the other was called the "Little Joker" (D543 in "Notebook"). The main difference between them was in the method of retaining the axle pin. The grip contour was not the same, grips were not interchangeable between the two types.

An often used pattern of cylinder decoration is pictured to right. This consists of a series of open diamonds created by a double line of punched zig-zags. This and other illustrations on this page are from **4650**. No finish remains, may have had gold wash initially.

Cylinder topstrap has a double scallop punched line on each side of sight groove. Shallow crescent engraved cuts border each scallop. The backstrap has geometrical lines bordered with punched hollow circles.

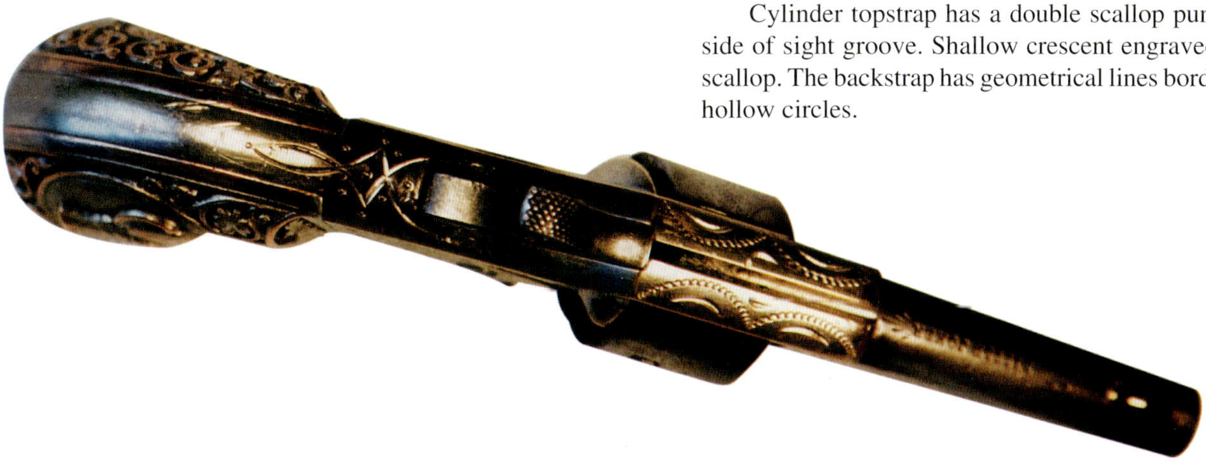

Marlin had metal grips decorated in high relief made by Francis DeGress of Bloomfield, New Jersey. These were in conjunction with his patent of 1874. They would of course have been offered at an additional cost.

The models XX Standard, XXX Standard, and O.K. had these grips portraying a lightly clad woman within the right panel.

The Little Joker's right panel employed the bust of a helmeted classical warrior.

Unlike the solid silver high relief grips made by Tiffany, the DeGress grips were made through an electro-type process wherein a thin layer of bronze was built up within the mold, and the body of the grip was filled with type-metal. The finished product would be nickel plated.

103+ Marston Superposed Barrels 103+

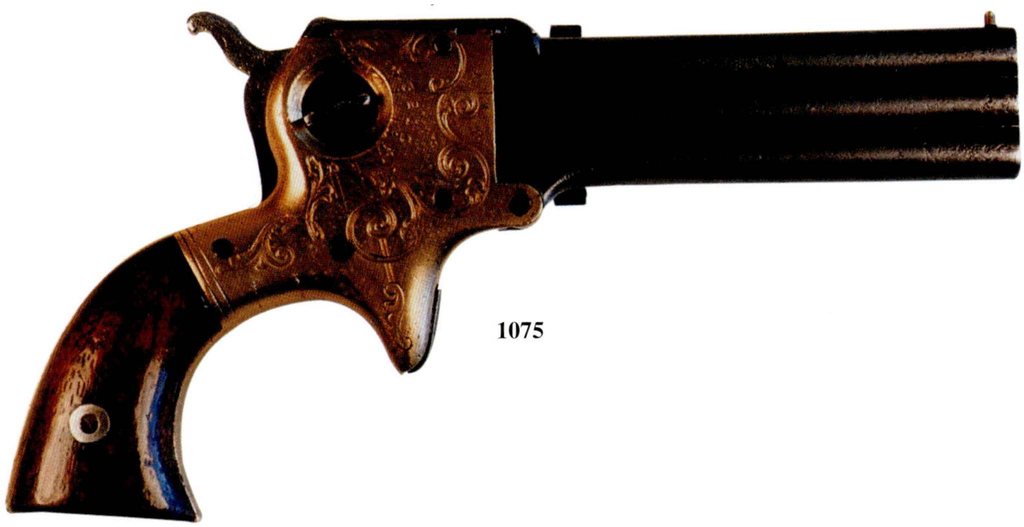

1075

A very open but well executed design is found on both sides of frame and on backstraps. The designs on both sides of frame is composed primarily of shaded foliate C-scroll engraving, and secondary punched five-pointed stars.

A two layer semi-sunburst pattern is cut into top of backstrap directly behind the hammer. This is very consistent throughout examples observed. At the base of the backstrap, a pattern composed of two ellipses, a flattened diamond, and another ellipse is found. This pattern varies slightly.

A single periphery line borders the backstrap, combines with double periphery line at back of frame.

Examples with very ornate and much fuller coverage have been seen, but they constitute the rare exception rather than normal method of decoration.

Unlike similar Marston pistols in larger calibre, the impressed inscription on left side of frame is always the same, viz:

WM. W. MARSTON	(top, curved line)
PATENTED	(straight horizontal line)
MAY 20, 1857	(straight horizontal line)
NEW YORK CITY	(bottom, curved line)

Left side of frame, pattern deviations:

(1) number and location of stars vary;

(2) scrolls in front may pass over or through screws. When passing through screws, the screw heads are engraved so as not to disrupt flow of scroll;

(3) a three-armed spray is usually present between branches of front and rear scrolls. Exceptions noted;

(4) straight lines usually connect the Cs of bottom frame scroll and top of trigger sheath scroll, these lines passing under the curvature of the frame scroll. Exceptions noted.

(5) distance between top center Cs varies, but Cs always connect with the horizontal double line.

(6) Other small differences may be seen, these attributed to the individual engraving and not an attempt to deviate from standard pattern.

1075

1082

WM W. MARSTON
PATENTED
NEW YORK CITY

603

WM W. MARSTON
PATENTED
NEW YORK CITY

808

As in other engraved areas, a basic design is adhered to. Number 1075 is a bit more ornate than others but the basic design is followed.

603

808

1075

1082

Louis Daniel Nimschke engraved several Marstons, but he was not responsible for the work on all pieces produced. He may have established the general pattern with lesser engravers performing the bulk of the work. Within Nimschke's record book of engravings executed by himself, six of the subject pieces are portrayed. Each is much more ornate, with much fuller coverage, than the norm; but the basic pattern is seen in all of them. Nimschke was born in Germany in 1832, migrated to this country c. 1850, spent most of his time in Brooklyn before his death in 1904. The majority of patterns he used were based on German patterns, as was the so-called "American" style of engraving also used by other engravers during the second half of the nineteenth century.

550+ Reid, Model One 550+

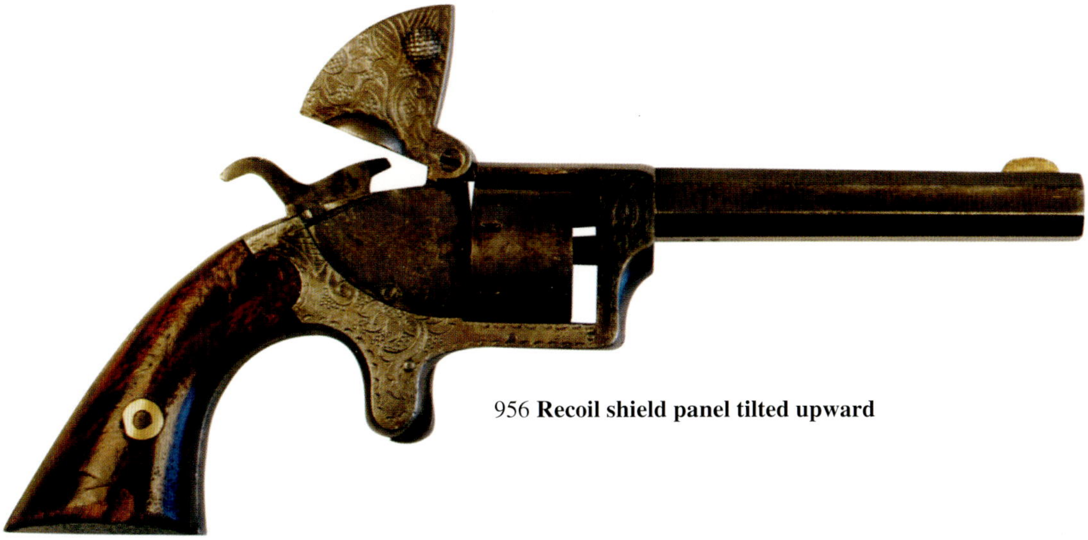

956 **Recoil shield panel tilted upward**

Prior to moving to Catskill, New York and the commencement of production of the "knuckleduster" (D105), James Reid produced revolvers in New York City.

His first model was made in .22 calibre only. Less than 900 of that Model One, in all three variations, were produced. Barrel inscriptions varied:

J. REID. NEW. YORK
JAMES P. FITCH. N.Y. PHOENIX REVOLVER
HOLEY MFG. CO. LAKEVILLE, CONN.
W. IRVING, 20.CLIFF.S.T.N.Y.
(or unmarked)

Only two examples of engraved pieces observed, both with the rare Holey Manufacturing Company inscription. To date, these are the only two such marked pieces in collections. Both have very similar engraving patterns.

Engraving is shallow, of mediocre craftsmanship. Other than the occasional foliate scroll, the overall coverage is unique to this model.

Barrel marking and decorated cylinder topstrap are shown on the illustration below.

This page illustrates variant views of serial number **956**

Backstrap

Note within above illustration, forward of cylinder, a series of semi-circles with hollow dots between lines. Might this be done to resemble holly leaves? This motif also appears on hump at rear of cylinder topstrap.

105+ The Engraving of Reid Knuckledusters 105+

4295

Upon close examination of several pistols, there appears to be a uniformity in the engraving and sequential patterns on frame sides and backstraps. This is not the case in other areas which will be dealt with later in this review.

A few pieces do have a definite difference in workmanship and patterns employed. They are serial numbers 4819 and 5892. Each of these is different from both the regular engraver and each other. Apparently at least two other engravers were involved for a brief time. This obviously is quite speculative, many engravers could have been responsible for the work throughout the company's existence.

Taylor Bowen, in his fine book relative to guns produced by James Reid, does mention a few engravers at the Catskill plant. One was James Reid, Jr., the other being Henry C. Thomas. Possibly others involved, but these two men probably did the majority of engraving on the knuckledusters.

Concerning the side of frame only, no two of the thirty examined are exactly the same. However, some major pattern changes are noticed; these have been listed under nine pattern headings. They are described and illustrated on following pages. With the exception of the two deviant patterns noted, each pattern falls within its own serial number range; thusly, it may be assumed the pattern changes were evolutionary in nature and not intermixed throughout production.

A great variety of designs were used on the topstraps. Pattern numbers have been assigned to basic motifs examined. With a few exceptions, the designs seem to have a sequential relationship with the progressive serial numbers. No examples have been observed that return to an earlier pattern, though some patterns are quite similar.

Decoration immediately behind hammer differs greatly, as does that on muzzle end. Some evolution of pattern types if sound behind hammer; not so at muzzle.

Engraving on recoil shields are very much alike throughout production. Rays radiating from edge to side of frame. Usually there are secondary smaller rays between the main arms.

Too few examples within this study to render results indicative of all patterns used. Undoubtedly there were others not covered herein.

15120

Obverse illustration is of the early period of manufacture. That pictured above represents engraving as found toward the end of production.

The periphery border line on side of frame and partially encircling finger hole ceased to be used between numbers 10528 and 13512. Silver plating was standard to number 10528, then nickel plating was introduced and used for remainder of production. There are a few early nickel plated pieces but that was the exception, not standard finish for the time.

Pattern # + DESCRIPTION, BARREL ENGRAVING		SERIAL #
31	Simple and sparse foliate shaded C-scrolls. No stippling. Periphery border line, interrupted in front of fingerhole. Foliate hammer screw. Plain trigger pin. Unadorned recoil shield. NB: This is the prototype model, made in New York City.	3454
32	Foliate shaded C-scrolls, solid pattern. Usually stippled background composed of small circles. Periphery border line; usually not open in front of fingerhole. Hammer screw area decorated; designs vary. Recoil shield has thin shading between rays; scalloped base.	3589 3675 4295 4631 5052 5331
33	Thin stemmed shaded foliate C-scrolls. No stippling. Periphery border line, open in front of fingerhole. Hammer screw and trigger pin areas not decorated. Three groups of serrated lines on recoil shield. NB: Engraving on all surfaces unique to this example; indicative of other than regular engraver.	4819
34	Light primarily foliate pattern with very little shading. Stippling in form of half circles in clusters. Leaf droops from hammer screw. Periphery border line; broken forward of fingerhole. Recoil shield has thin shading between rays; plain base. NB: Indicative of other than regular engraver.	5892
35	Shaded foliate C-scrolls with hint of flower petals at rear. Stippled background; some stippling in form of cluster of berries extending beyond engraving. Periphery border line; open at front of fingerhole. Recoil shield has thin shading between rays; plain base.	5957 5967 6728
36	Shaded foliate C-scrolls. Stippled background in form of clusters of small circles with open spaces between clusters. May have foliate pattern between hammer screw and trigger pin. Recoil shield has thin shading between rays; plain base.	9569 9931
37	Foliate shaded C-scrolls; solid pattern with much stippling. Foliate pattern between hammer screw and trigger pin. Periphery border line, not open in front of fingerhole. Recoil shield has thin shading between rays, plain base.	9943 10528

Pattern # + DESCRIPTION, BARREL ENGRAVING		SERIAL #
38	Very tight shaded foliate C-scrolls. Large blocks of stippling within and outside of main pattern. No periphery border lines. May or may not have engraving between hammer screw and trigger pin; when present, usually profuse. Recoil shield has thick rays, some with background shading; plain base.	13512 13796 14384 15120 15121 15923 16463
39	Shaded foliate C-scrolls, flower petals near sideplate screw. Solid heavy stippling within basic pattern. Foliate pattern on and above trigger pin. No periphery border line (exception: number 17206 which is later illustrated.) Recoil shield has thin shading between rays; plain base.	17206 17474 17557 17620

3454
31

17474
39

Pattern # + DESCRIPTION, BARREL ENGRAVING		SERIAL #
40	Thin straight periphery border line. Groove: Shaded scallops, small thin line above and between each.	3454
41	No periphery border line. Groove: Thin straight line, shaded tulip in center.	4819
42	Series of interrupted scallops on side edges; front and back edges plain. Groove: three curves, nearly semi-circular, slight shading above each curve.	5967
43	Broken periphery border line on each side. Series of curved lines at front and back. Groove: series of elongated diamonds.	5052 5892 5957
44	Series of shaded scallops on sides; curved lines front and back. Foliate spray in center of each side of groove. Groove: series of plain scallops. NB: Gold washed and red lacquor in deep cuts. Only example examined with full foliate pattern on topstrap.	5331
45	Periphery border lines on each side either zig-zag or straight. Thin curved line at front and back. Groove: fairly large shaded scallops, straight thin lines above and between each scallop - or series of zig-zag lines bordering each side of groove.	3589 3675 4295 4631
46	Series of interrupted scallops on side edges, nothing on front or back edges. Groove: four curves on each side, slight shading above each curve. NB: similar to Pattern 42, but curves cut much deeper with flatter radius.	10528 13796
47	Series of solid ellipses on sides, front and back edges plain. Groove: Four shallow flat curves, each connected by cluster of three or four punched hollow circles; shading above each curve.	13512 14384 15120 15121 15923

Pattern # + DESCRIPTION, BARREL ENGRAVING		SERIAL #
48	Series of solid ellipses on sides; front and back edges plain. Groove: wavy line each side, curved lines as offshoots, each with cluster of three punched hollow circles at end; creates appearance of vine with leaves and buds.	16463
49	Series of scallops (may be solid or shaded) separated by punched hollow circles, on sides. Front and back edges plain. Groove: three or four shallow curves or small radius, each connected by clusters of punched hollow circles. Shading above each curve. NB: similar to Pattern 47, difference being side decoration.	17206 17620
50	Series of solid ellipses, separated by punched hollow circles, on sides. Front and back edges plain. Groove: wavy line each side, resembles vine with protruding leaves. Slight shading above each "leaf".	17474
51	No periphery border pattern. Groove: several rows of solid ellipses, some with puched hollow circles at top. Circles randomly placed.	17557

3589
45

4295
45

Pattern # + DESCRIPTION, BARREL ENGRAVING		SERIAL #
70	Flat sided shield, slight point at top. Base constricted, flows into diamond, then teardrop. Continuous checkering through all areas. Double border line.	3454
71	Shield with concave top; two ellipses at bottom. Continuous checkering through all areas. Double border line. NB: inscription forward of shield.	3589
72	Shield with slight point at top; two ellipses below shield. Continuous checkering through all areas. Double border line, small loop at top.	3675
73	Checkered ellipse. Single border line touches checkering. Second border line continues above and below checkering, each end terminating in an ellipse followed by a diamond.	4295
74	Checkered oval. Single border line touches checkering. Second border line continues above and below checkering, each end terminating in a small circle.	4819
75	Checkered ellipse. Same as Pattern 73 except outer border line terminates at both ends with small loop; no diamond.	5052 5331
76	Checkered shield, slight point at top. Single border line touches checkering. Second border line has loop at top, small ellipse at base.	6728
77	Checkered shield, curved sides, slight point at top. Single border line touches checkering. Second border line has two elliptical loops at bottom.	15923
78	Acorn shaped checkered area, pointed at base. Single border line touches checkering. Second border line ends in loop at top which touches checkering, and double loop at bottom which does not touch checkering.	17206 17557

These are the basic patterns. Pattern 78 shown twice in illustrations to show slight differences; such will appear within all patterns.

3454
40

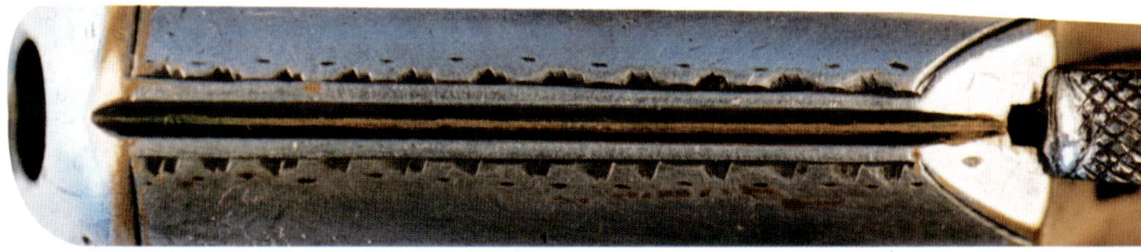

4819
41

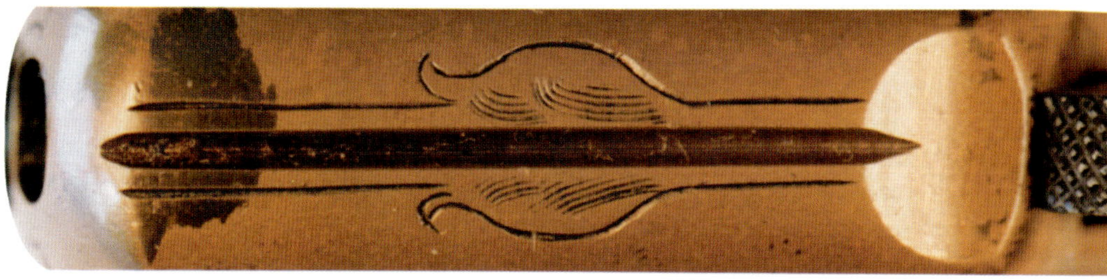

5967
42

5957
43

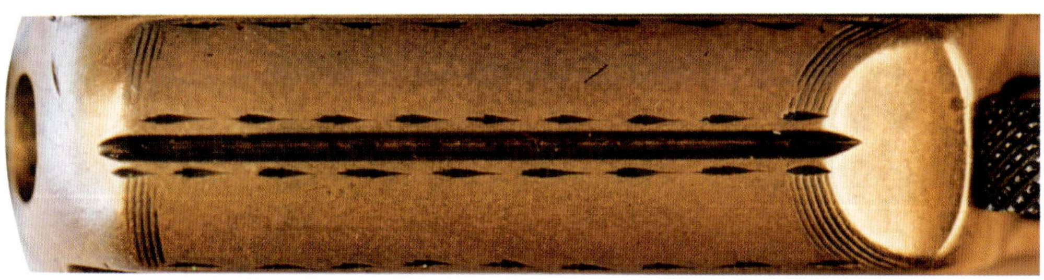

5331
44

3675
45

10528
46

14384
47

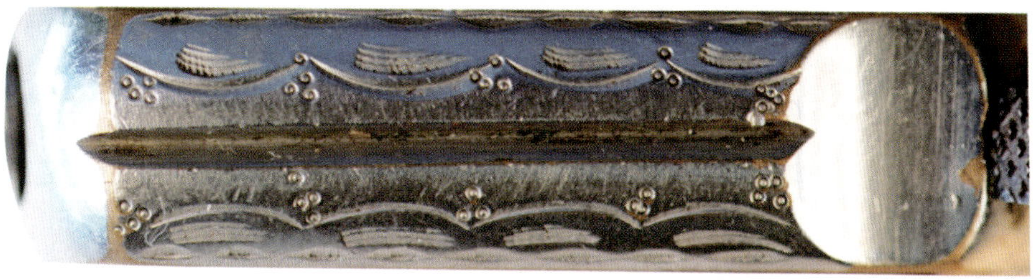

16463
48

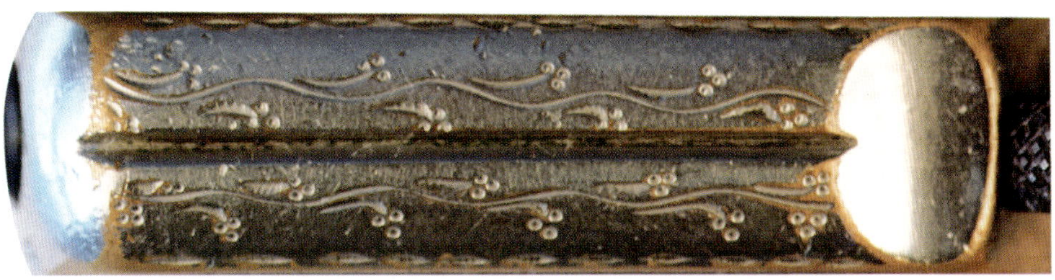

17620
49

17474
50

17557
51

3454
70

3589
71

3675
72

4295
73

4819
74

5052
75

6728
76

15923
77

17206
78

17557
78

Ten representative engraving patterns are pictured below. The engraving found immediately behind hammers have no progressive sequence of development as seen in other areas.

There is also no consistency relative to matching of patterns for back of backstrap to top of backstrap. However, during the last few years of production, those two areas very often match each other through the employment of radiating rays as seen on **15121** and **17557**.

3454

3675

4819

5331

5967

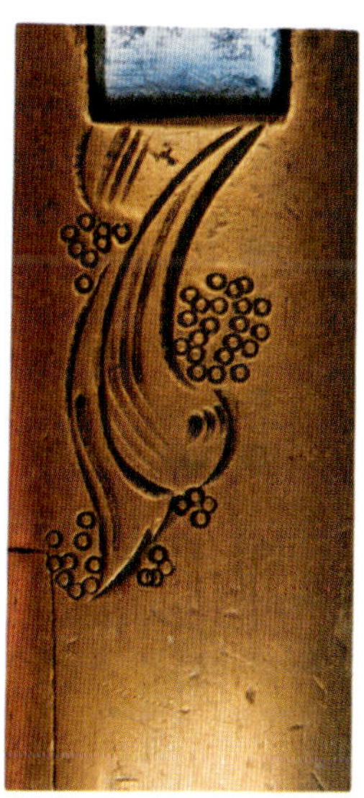

9569

10528

15121

17474

17557

The following four pages illustrate muzzle ends of knucklers in sequential serial number order. Basic patterns are covered, plus a few unusual ones. The last two pages are clusters showing very little in change, they are included to indicate uniformity toward end of production.

Comments, by serial number, follow:

3454. New York City prototype. No decoration.

3589. Note periphery leaves, and petals surrounding axle pin.

4295. Periphery decoration continues upward into topstrap area.

4819. Unusual pattern, as is all engraving on this gun.

5052 through **6728.** Conventional radiating rays, periphery straight border line which does not extend into topstrap area.

9931 and **9943.** Conventional pattern without periphery border line. Both of these guns are of the transition period when safety was eliminated; both have fillers in safety slots in bottomstrap.

10528 through **14384.** Conventional pattern without periphery border line. Slight differences in secondary rays.

16464. Conventional pattern, but with clusters of punched circles at tips of some rays.

17557. Conventional pattern but with concentric periphery border lines; some straight, some wavy.

17206 and **17620.** Conventional.

17557 and **17620** have iron frames. Toward end of production, patterns remained basically the same whether frames were of iron or brass.

4631

3454
4295

3589
4819

5052
5957

5892
6728

9931

9943

10528

13512

13796

14384

17557

17206

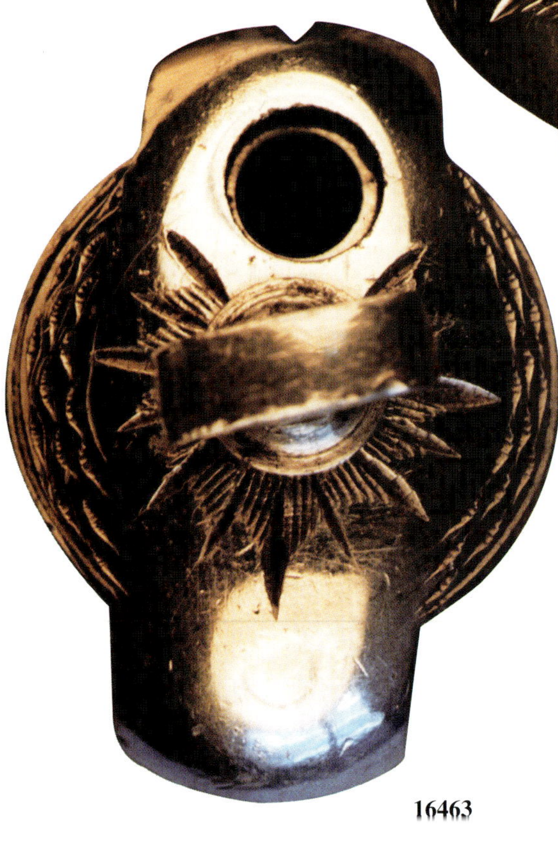

16463

17620

3589
32

4819
33

5892
34

5967
35

9569
36

10528
37

15121
38

17206
39

105+ Reid Knuckledusters, Epilogue 105+

Though this study is confined to handguns of .22 calibre, a few of the larger calibres are included. Reason: to illustrate the continuation and expansion of engraving patterns associated with Reid knuckledusters.

This page contains pictures of the second variation of the .32 calibre Knuckler with barrel, serial number **13374**. Engraving is consistent with standard patterns, but the flat topstrap presented an area where a fuller pattern could be used - as was.

Left: Top view of topstrap.

Right: Backstrap engraving. Patterns as used on knucklers within same serial number range; with or without barrels.

17777

With departure from the conentional Knuckler pattern, this all metal revolver in .41 calibre rim fire, was produced in limited quantity in the year 1882.

The scalloped decoration on edges of topstrap, and the checkered area on rear of backstrap, follow the same patterns as on the conventional knucklers. There is no engraving on recoil shields as is always found on the knuckler. Sides of frame and grip area follow the normal foliate C-scroll pattern with stippled background as typical of the knucklers.

A new element is found on side of frame forward of cylinder, that being a rectangular square within quilted diamond pattern with hollow circles within alternating diamonds.

232 Remington, Vest Pocket 232

C301

This model was produced from the mid 1860s to 1888. Engraved pieces could be had from the factory, though the bulk of such pieces were probably decorated by local engravers for large retail sporting goods firms.

The common foliate C-scroll pattern is frequently seen, usually with stippled background. A few have a rather distinctive pattern with double diagonal lines on sides of frame, together with C-scrolls on each side of diagonals. This was probably the work of one engraver.

Example pictured is housed in gutta percha case (originally made for a Sharps pepperbox). Pearl handles on accessories.

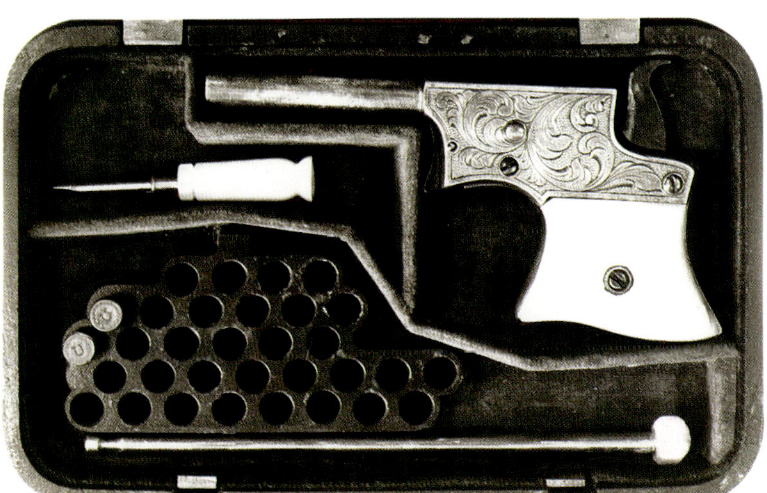

110 Remington, Elliot Pepperbox 110

Illustrations are of serial number **7188**. Though engraved Elliots were usually silver or nickel plated, this example has original blue/grey finish.

Production probably began in 1863, was discontinued in the late 1870s. Engraved specimen of the earlier Zig-Zag model exist; unfortunately, none available for illustration herein.

Backstrap engraving.

Buttstrap engraving.

561+ Rupertus 561+

3446

Using the machinery of the Tryon company of Philadelphia, Pennsylvania, Jakob Rupertus produced several types of firearms. Herein illustrated is one of many variants of his EMPIRE PATENT revolver. This would fall into the "suicide special" class as first appeared in the 1870s. It is included to provide an example of the poor crude decoration associated with this classification.

A shallow C-scroll pattern appears on the sides of frame; the remainder of decoration being punched circles, periphery lines, and wavy line as seen on barrel. Obviously the quality exemplified by the engravers of the preceding decades is not present.

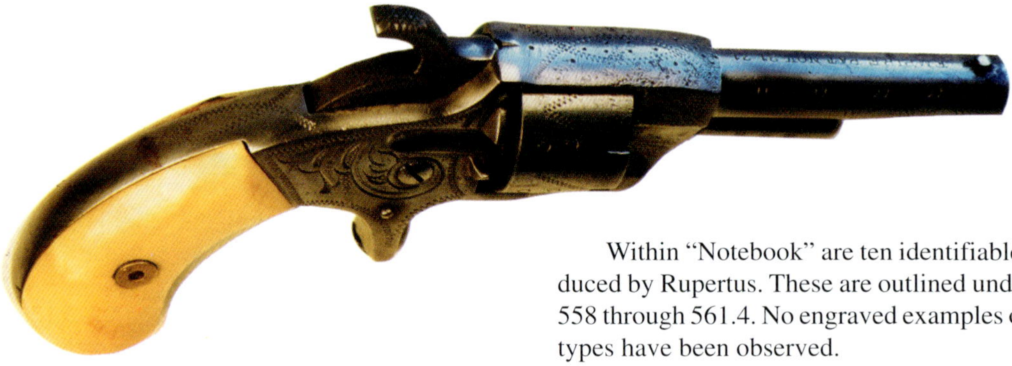

Within "Notebook" are ten identifiable types of revolvers produced by Rupertus. These are outlined under Description Numbers 558 through 561.4. No engraved examples of the earlier better made types have been observed.

115+ Sharps Pepperbox 115+

295

Pictured above is the variation with inscription on lower edge of barrel cluster (D115). The engraving on this piece is not typical of the standard patterns employed. The use of the rosette is more emphasized and the background is much plainer than the usual. Refer to page nine for enlarged view.

The rosette was used on Sharps handguns more frequently than on products of other makers. One large rosette encircled the hammer pivot screw on most pieces, later models having company inscription encircling the rosette. Often the screw on side of frame extension is also surrounded by a rosette.

Excellent coverage of Sharps' engraving patterns, with many fine illustrations, may be found in Frank M. Sellers' fine book entitled "Sharps Firearms". This work is very detailed, several outstanding pieces are pictured, cannot be more highly recommended.

As with engraved pieces of contemporary makers, Sharps relied heavily on the shaded foliate C-scroll pattern. The scrolls were mostly on sides of frame, some examples continued the pattern to include sides of standing breech. Stippling was often used in background, as less frequently was an area of quilted diamonds. Periphery lines, often wavy with series of solid scallops, are frequently seen.

The barrel cluster is usually devoid of engraving. When engraving is present, normally in the form of foliate C-scrolls at rear of each tube, and small geometric patterns near muzzle. Examples have been seen with full geometric designs and punchwork in flutes of tubes; these were probably not of Sharps creation.

3430

The buttstraps on earlier examples often have engraved periphery double lines, usually curved; no engraving on crosspiece. Later production, when engraved, may have sprays with checkered backgrounds, and radiating lines on crosspiece.

A double periphery line, sometimes wavy, usually borders the sides and top of standing breech. This area may have geometric or foliate engraving, sometimes intermixed.

Note C-scroll above quilted pattern on above. Also note small rosette encircling screw on side of frame extension. Another example of inscription on bottom of barrel cluster rather than on side of frame.

Finish usually of silver frame with blued barrel cluster. Some had both frame and barrel cluster silver plated. Rarely one will be seen with nickel plating.

Serial number **36511**. Exemplifies engraving of the more often encountered type as used in later production. Note tight coverage, much stippling in background. Rosette encircling hammer screw still being used.

35989

Above is an example of shell pattern in area of hammer pivot screw.

To right:
Englarged view of 35989.
Most encountered form of Sharps' engraving.

Most engraved examples have a shell at top of backstrap, immediately behind hammer. Different forms of secondary decoration are found in this area. A modified form of half rosette will also be seen.

The backstraps on many of the earlier pieces have a periphery border line, then clusters of leaves at top and bottom. A blank space is left in the center, presumably for inscription if desired by purchaser. Note **295** and **3430**, both pictured below, have opposing patterns, one from right, the other from left.

35989

295

3440

An example of later backstrap decoration is seen on **35989**. Scalloped border, ellipses and diamonds in middle area.

572+ Smith & Wesson 572+

45533

The **Pattern Numbers** relate to the Smith & Wesson Model One Third Variation only. No engraved First or Second Variations were available for comparison. To view engraved examples of those issues, refer to book of William Locke's Collection, pp. 361, and Roy Jinks' "Artistry In Arms."

There appears to be more consistency within engraving patterns by Smith & Wesson than with other producers. Quality is also more uniform. The majority of those examined have a nickel finish, a few with blued finish, and none with silver or gold wash - though the latter may exist.

The patterns are very similar to other gunmakers of the period. Other than zig-zag lines and small circles, all patterns are engraved with no punch or stamp work as was so common on the later "suicide special" type revolvers.

When a revolver is engraved, all areas have some type of decoration. No partially engraved pieces have been seen as was the case with some other gunmakers.

Toward the end of production, the periphery border line on sides of frame was discontinued. This is also true of the Whitneyville and Reid products.

The right side of frame always has the same general pattern as does the left side.

Though some revolvers were decorated by engravers working directly for the company, most were done through outside contracts either by Smith & Wesson or, most often, by large firearms distributors or retailers.

Fine examples of engraved pieces, along with backgrounds of principal engravers for Smith & Wesson, will be found in "Artistry In Arms", by Roy C. Jinks. A few notes relative to known engravers for Smith & Wesson, from the Jinks' work, follow.

Richard Bates Inshaw, of Springfield, was the first to engrave for Smith & Wesson. A fine illustration of his work appears in text mentioned. Cost of engraving was $2.00 on a First Model, First Issue revolver.

Other engravers who did work for Smith & Wesson in their own quarters were S.T. Merritt, F.W. Martin, E.A. Timme, and P.S. Yendell. Their work was not confined to Smith & Wesson, other firearms makers also had pieces done by them.

Gustave Young learned the art of engraving in Germany. He migrated to this country in 1846. Located in Hartford, Connecticut in 1852, did work for Colt and Smith & Wesson. He moved to Springfield in 1868 and was employed by Smith & Wesson for the remainder of his engraving career. Two sons, both engravers, worked with the company for several years. Their activities were not confined to engraving alone, die making and related work probably consumed much more time than the engraving of handguns.

The famous engraver, Louis D. Nimschke, also engraved Smith & Wesson handguns, though the majority of them were for firearms dealers in the New York City area.

Again, the reader is encouraged to become familiar with the contents of "Artistry In Arms".

Recapitulation:
Smith & Wesson had local engravers decorate firearms, this being through outside contracts.

Their work was not confined to Smith & Wesson handguns, they also did work for other firearms producers.

A few engravers were employed full time by Smith & Wesson; activities were not confined to engraving guns.

Probably most of the engraved handguns were the result of outside contracts with engravers by major firearms dealers and distributors.

Pattern # + DESCRIPTION		SERIAL #
1	Foliate shaded C-scrolls, stippled background within scrolls. Wide coverage in center only, no periphery line. Serrated band encircles muzzle; Side Marking variant has band of scallops with triangles encircling muzzle.	42167 50749
2	Foliate shaded C-scrolls with stippled background. Large coverage extending rearward from muzzle. Smaller coverage extending forward from rear of barrel.	77188
3	Shaded leaves front and rear. Diagonally rearward sloping scales throughout center. Top and bottom lines with small scallops, dots between scallops.	82573
4	Series of squares with clipped corners, single circles within each; serrated lines between squares. Series of shaded scallops above and below squares with line between each scallop. Sprays fore and aft of main pattern.	93128
5	Shaded leaves front and rear. Wavy zig-zag line in center, solid petals by each curve.	107760
6	Series of squares with clipped corners, concentric circles within each. Serrated lines between squares, series of elliptical lines separated by circles above and below squares. Foliate C-scroll fore and aft of main pattern.	122354 101981
7	Foliate shaded C-scrolls, long thin pattern, central area only. Stippled background within scrolls. No band near muzzle.	45533 76952
8	Two flower petals in center, then vines with leaves to front and rear. Plain near muzzle. Stippling within petals and leaves.	99841
9	Double periphery line above and below **Side Marking** inscription. Shaded scallops on both sides of periphery lines, small solid diamonds between scallops. Ellipse with stippled center fore and aft of inscription. (Left Side)	76852
10	Series of scallops above and below **Side Marking** inscription. Single petal fore and aft of inscription. Hollow dots on both sides of junctures of scallops. Crescent outside each scallop.	 76952
11	Straight periphery double line with stippled ellipses fore and aft. Series of shaded scallops outside horizontal periphery lines, diamonds between scallops. Series of shaded oval blocks between periphery lines, stippled background. Similar to patterns 4 and 6. **Side Marking** variant, right side only.	 76852

Pattern # + DESCRIPTION		SERIAL #
11	None. Plain. Front of topstrap plain.	42167
12	Wavy line bordering inscription. Topstrap plain (45533). Two petals each side at rear, stippling in center between petals. Three radiating lines on front of topstrap (77188).	45533 77188
13	Series of elliptical cuts separated by small circles border inscription. Double curved line at rear. Quilted diamonds on front of topstrap.	82573
14	Wavy punched zig-zag line bordering inscription. Double curved line at rear. Radiating lines on topstrap.	93128
15	Double wavy line bordering inscription. Double curved line at rear. Front of topstrap stippled.	107760
16	Series of elliptical cuts separated by small circles border inscription. Curved line of small ellipses at rear. Shaded petals at front of topstrap.	122354
17	Series of shaded scallops border each side of inscription. Small solid triangle between junctures of each scallop. Double curved line at rear. Radiating lines on front of each side of topstrap.	101981
18	Thick serrated wavy line borders each side of inscription. Double curved line at rear; serrations between rear line and edge, series of thin ellipses out into serrations. Petals on each side of front of topstrap.	99841
19	Foliate C-scroll at rear, shaded with stippled background. Four ellipses, descending size toward muzzle, bordering punched hollow circles. No periphery border lines. Three radiating lines on each side of forward area of cylinder topstrap. (Side marking variant, D573)	76952
20	Five deeply cut lines in spray pattern at rear; unconnected series of shaded scallops on each side of rib, small shaded diamonds between scallops. V and ellipse at front. Diamond patterns within periphery lines on each side of forward area of cylinder topstrap. (Side marking variant, D573)	76852
21	Two shaded leaves at rear; series of connected scallops on each side of rib, small shaded diamonds between scallops. Straight periphery line at front. Three radiating lines on each side of forward area of cylinder topstrap. (Side marking variant, D573)	50749

Pattern # + DESCRIPTION		SERIAL #
30	Downward sloping C-scrolls, starting at front. Main scroll encircles screw. Wavy line and scallops on topstrap. Stippled background. Periphery border line. Medium shading.	42167
31	Upward sloping foliate C-scrolls. Main scroll above and to rear of screw. Stippled background. Periphery border line. Slight shading. Small foliate cluster near hinge; shaded scallops on topstrap.	45533
32	Downward sloping foliate C-scrolls, starting at rear. Main scroll near rear. Stippled background. Periphery border line. Medium shading. Small foliate cluster near hinge screw; scallops on topstrap. Hollow circles within series of diamonds on topstrap.	77188 122354 (77188) (122354)
33	Upward sloping foliate C-scrolls. Periphery border line. Main scroll above and to rear of screw. Stippled background. Large foliate cluster near hinge screw; scallops on topstrap. Foliate cluster near hinge screw; double wavy line on topstrap. Foliate cluster near hinge; shaded scallops on topstrap.	82573 93128 (82573) (93128) (76852)
34	Upward sloping foliate C-scrolls. No periphery border line. Main scroll above and to rear of screw. Stippled background. Large foliate cluster near hinge screw; wavy line on topstrap.	107760
35	Upward sloping shaded foliate C-scrolls, more scrolls than leaves. Stippled background. Periphery border line. Main scroll behind screw, petals slope downward. Small leaves near hinge screw; shaded scallops on topstrap.	101981 76952
36	Full blossoms and assorted flowers. No shading or stippling. Double periphery line, foliate clusters in front corners. Small foliate cluster near hinge screw; ellipses separated by hollow circles on topstrap.	99841
37	Upward sloping foliate c-scroll, one major branch passes over main branch near top rear. Periphery border line. Stippled background. Wavy line on topstrap, cluster of hollow circles near hinge screw.	50749

NB: Topstrap as used above indicates side of cylinder topstrap.
When serial number in parenthesis, comment on that line refers to that specific revolver only, not all within Pattern Number designation.

Pattern # + DESCRIPTION		SERIAL #
60	Three sets of three petals, shaded, beginning at front. Stippling between and above petals. No design at rear.	42167
61	Tulip, beginning at front, shaded; may have series of hollow punched circles rising from each petal. Rear may be plain or have series of shaded scallops.	45533 76952
62	Shaded leaves pointing rearward. Series of small scallops with shading near rear edge and to rear of locking slots. Series of smaller triangles pointing rearward between scallops, some with scallops but no triangles.	101981 77188 50749 76852 82573
63	Spray with seven branches starting at front. No design at rear.	93128
64	Shaded leaves pointing rearward. No design at rear.	107760
65	Shaded leaves pointing rearward. Double wavy line near edge and to rear of locking slots.	122354
66	Side view of flower with petals on stems, between flutes. Different flowers, some with stippled clusters as berries. Two lines encircle rear area, serrations between edge and rearmost line.	99841

50749
62

76952
61

76852
62

Pattern # + DESCRIPTION		SERIAL #
70	Sunburst with small circles behind hammer. Periphery lines. Small shaded leaf patterns at top and bottom of strap. Two circles in band immediately above foliate pattern.	42167
71	Foliate c-scrolls, stippled and shaded. No periphery line. Begins rear of hammer, extends halfway down backstrap.	45533
72	Foliate shaded c-scroll rear of hammer, stippled background. Periphery line full length. Ellipse, diamond, double line ellipse. Small punched circles near juncture of curves.	77188
73	Foliate shaded c-scrolls. Stippled background, extends forward along sides of hammer. Double line ellipse, diamond, ellipse, ellipse, two teardrops.	82573 101981
	Variant with initial behind hammer, remainder of pattern same.	76852
74	Sunburst with seven rays behind hammer. Double-line ellipse, diamond, ellipse, ellipse, teardrop. Bordering circles.	93128
75	Spray pointing toward hammer; seven branches with shading. Double-line quilted triangle, ellipse, diamond, ellipse, ellipse, teardrop. Bordering circles near junctures.	107760
76	Shaded foliate c-scroll behind hammer. Elliptical cuts separated by small solid dots from right side of hammer to midpoint in strap; then double-line ellipse, diamond, ellipse.	122354
77	Shaded shell with dot within each segment, base trails into concentric circles. Petals with bud at base, then elongated serrated diamond. Double periphery border line. Area near sides of hammer has series of solid thin ellipses separated by solid dots which are engraved, not punched.	99841
78	Foliate c-scrolls at top, periphery border line ending in double-line ellipse at base. Inner line of shaded scallops, solid triangles between scallops. Name engraved in middle area.	76952
79	Sunburst behind hammer with eight deep rays, secondary rays composed of lines if hollow circles between each primary ray. Series of scallops beside hammer, small solid triangles between scallops. Below sunburst is double-line ellipse, diamond, ellipse, diamond, ellipse. Hollow punched circles border geometric designs. Similar to Pattern 74.	50749

Pattern # + DESCRIPTION (Top, Center, Bottom)		SERIAL #
90	Plain - Foliate - Plain.	42167 45533 122354
91	Foliate - Foliate - Plain.	77188
92	Foliate - Quilted diamonds, circles in each - Plain.	82573
93	Plain with double line at rear - Sunburst with double line at rear and double line of small triangles above and below - Plain.	93128 101981
	Variant with extension of side barrel pattern at top.	76852
94	Foliate - Sunburst - Plain.	107760
95	Plain; no line at rear - Sunburst; small triangles above and below - Plain.	99841
96	Extension of side barrel pattern. Diamond, half diamond, bordering circles and crescents. Plain.	76952
97	Plain. Ellipse and half diamond bordered by small punched circles. Plain.	50749

101981

Side	**35**
Cylinder	**62**
Forward	**93**

45533

Side 31
Cylinder 61
Forward 90

82573

Side 33
Cylinder 62
Forward 92

77188

Side 32
Cylinder 62
Forward 91

122354

Side 32
Cylinder 65
Forward 90

107760

Side	**34**
Cylinder	**64**
Forward	**94**

93128

Side	**33**
Cylinder	**63**
Forward	**93**

42167

Side 30
Cylinder 60
Forward 90

99841

Side 36
Cylinder 66
Forward 95

82573

Side 3
Forward 92

122354

Side 6
Forward 90

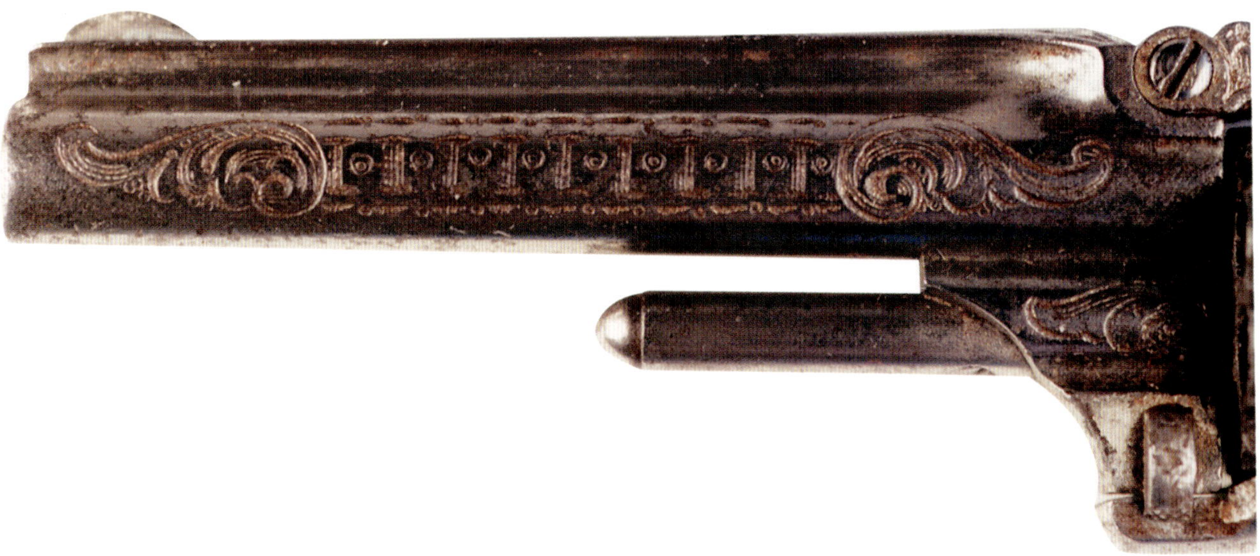

93128

Side 4
Forward 93

42167

Side 1
Forward 90

45533

Side 7
Forward 90

77188

Side 2
Forward 91

107760

Side 5
Forward 94

Tops of barrels, **Side Marking** variation. Pattern numbers are described on page 71.

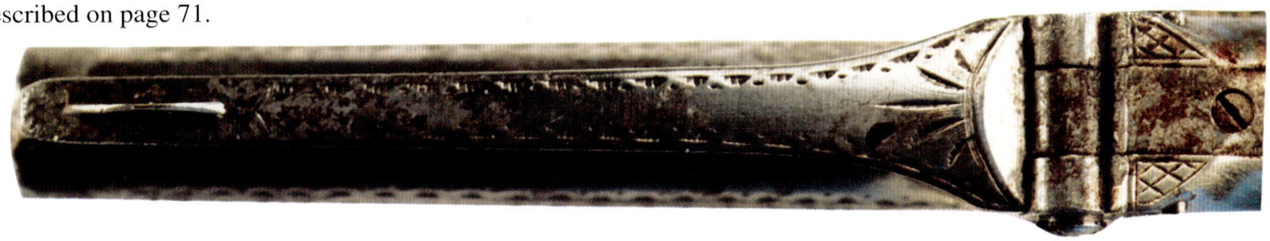

76852

Top 20

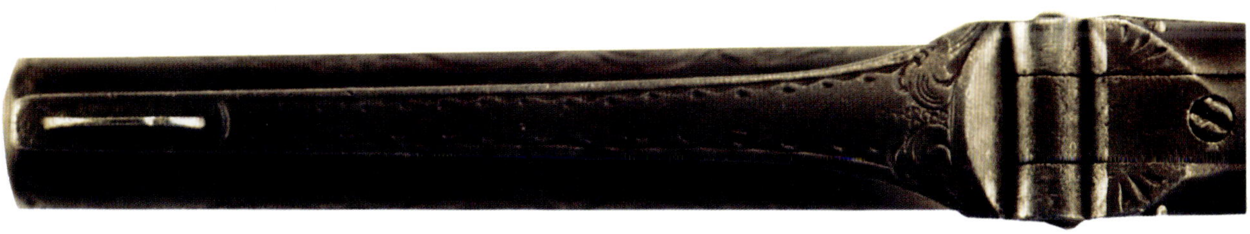

50749

Top 21

76952

Top 19

76852

Side **9**
Forward **93**

76952

Side **10**
Forward **96**

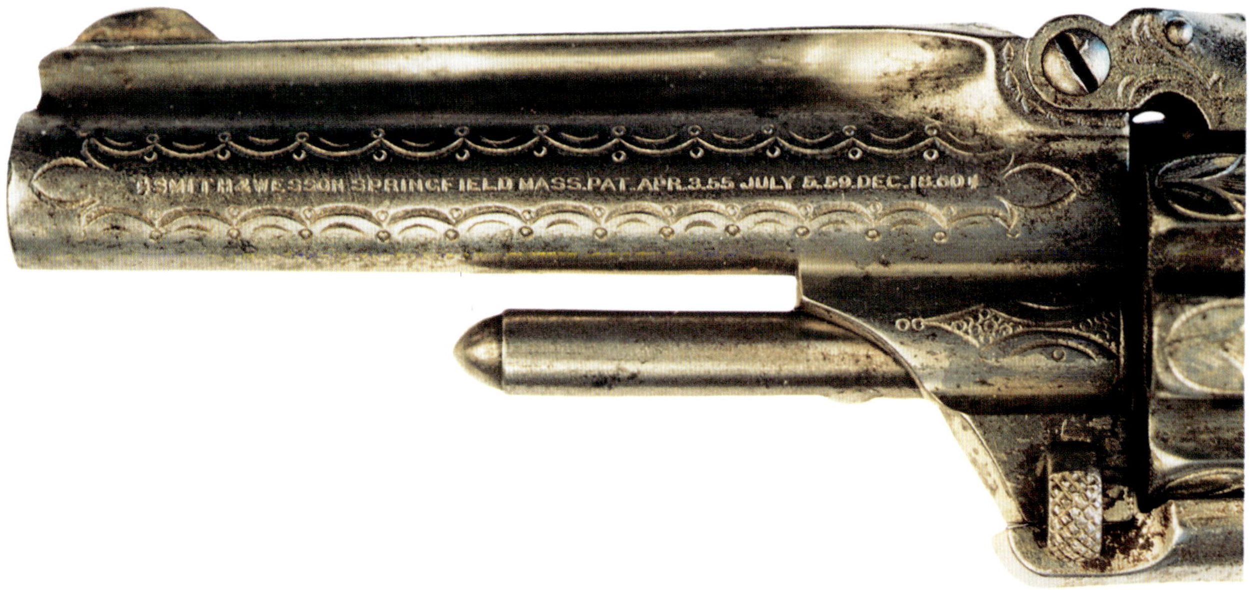

50749

Side **1**
Forward **97**

Note how engraving covers corporate inscription on **50749**. Also note design encircling barrel near muzzle.

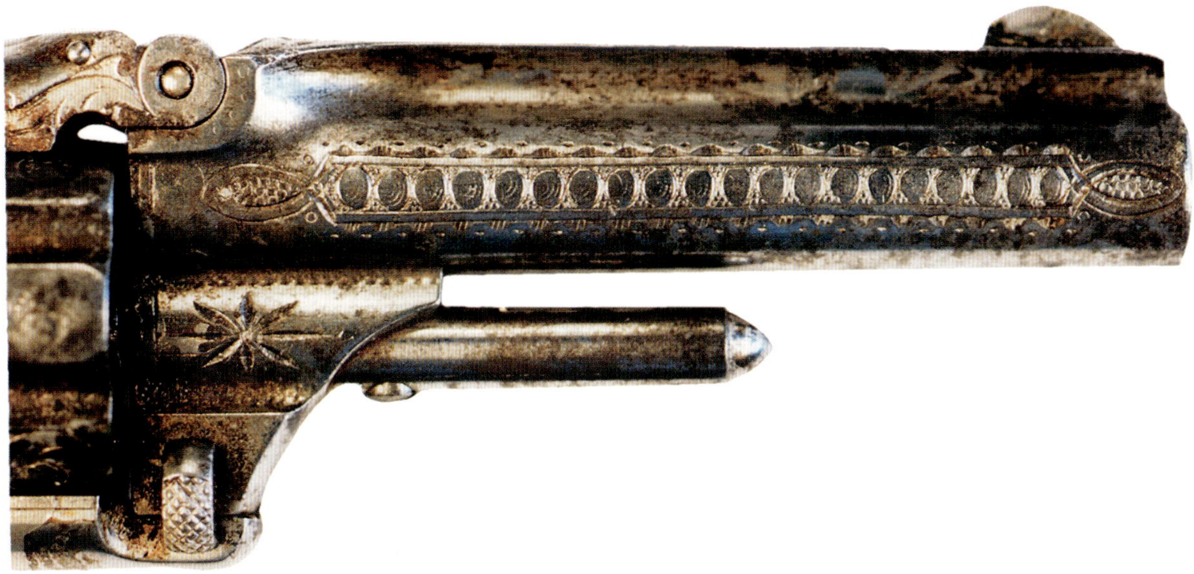

76852

Side **11**
Forward **93**

76952

Side **10**
Forward **96**

50749

Side **1**
Forward **97**

42167
70

77188
72

82573
73

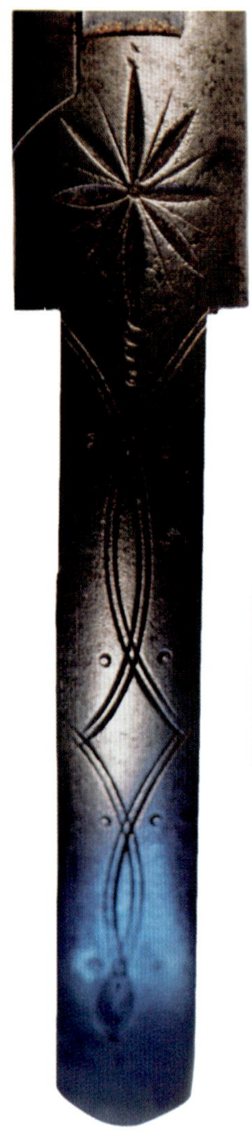

93128
74

107760
75

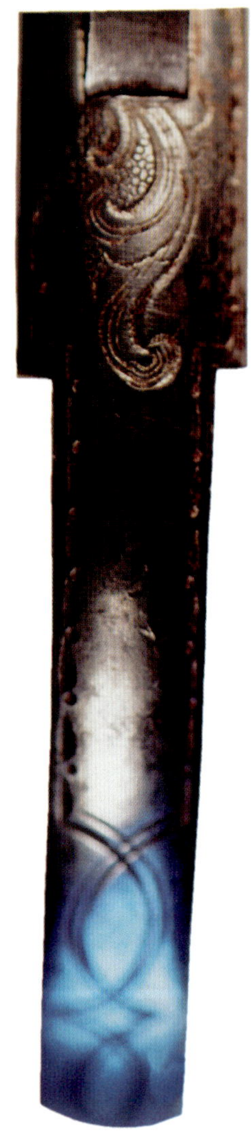

122354
76

45533
71

101981
73

76852
73

99841
77

76952
78

50749
79

99841
18

101981
17

82573
13

107760
15

45533
12

77188
12

A scarce variation of the Third Issue of the Model One is that with the corporate marking on left side of barrel rather then on top of barrel rib. Examples of this variation have shorter barrels than the regular model, they are 69mm long, or nearly 2 3/4 inches.

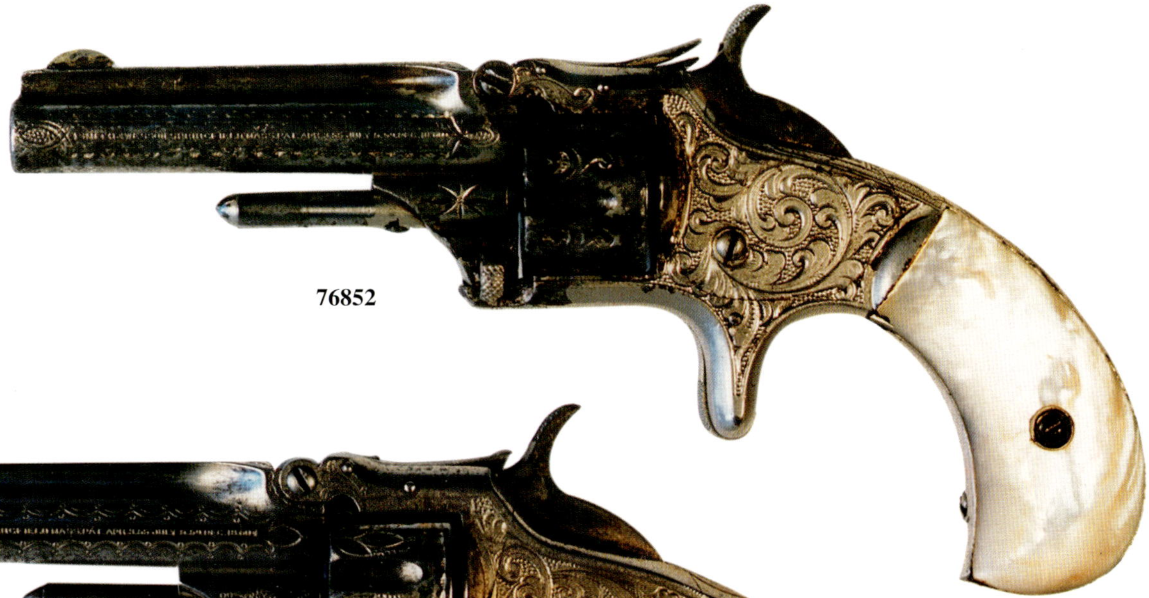

76852

76952

50749

L. Richard Littlefield has kindly supplied examples with serial numbers **50749** and **76952** for inclusion in this study. Engraved "side marking" pieces are extremely rare.

Other than top and side of barrel, pattern numbers listed under **572+** will be used. Some illustrations of the side marking variant will be found under that heading.

76852

Side 33
Cylinder 62
Forward 93

76952

Side 35
Cylinder 61
Forward 96

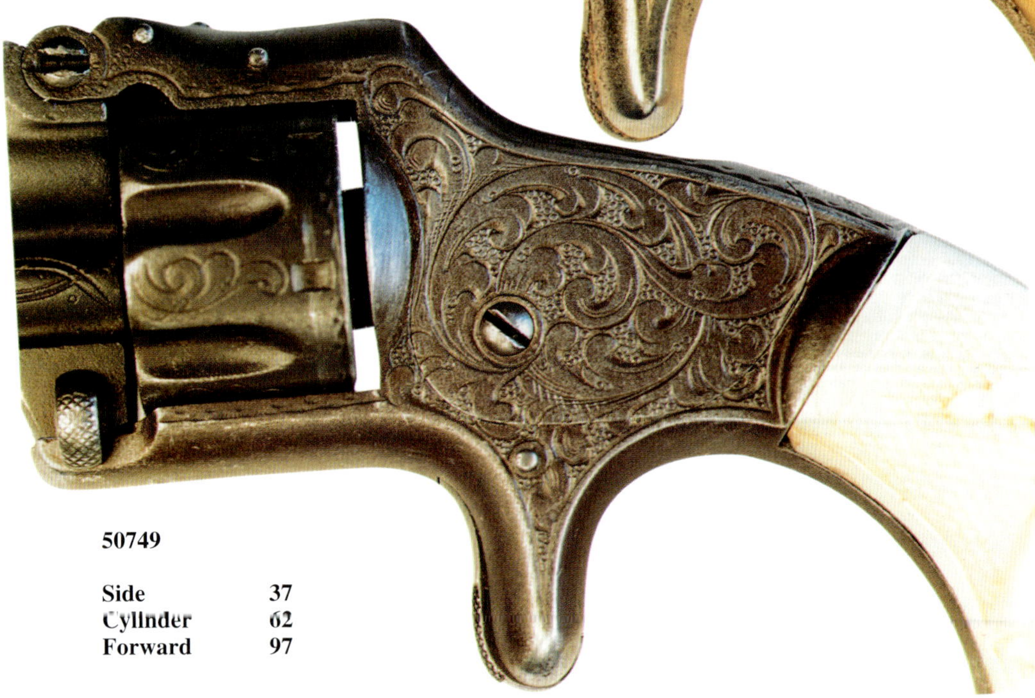

50749

Side 37
Cylinder 62
Forward 97

234 Stafford 234

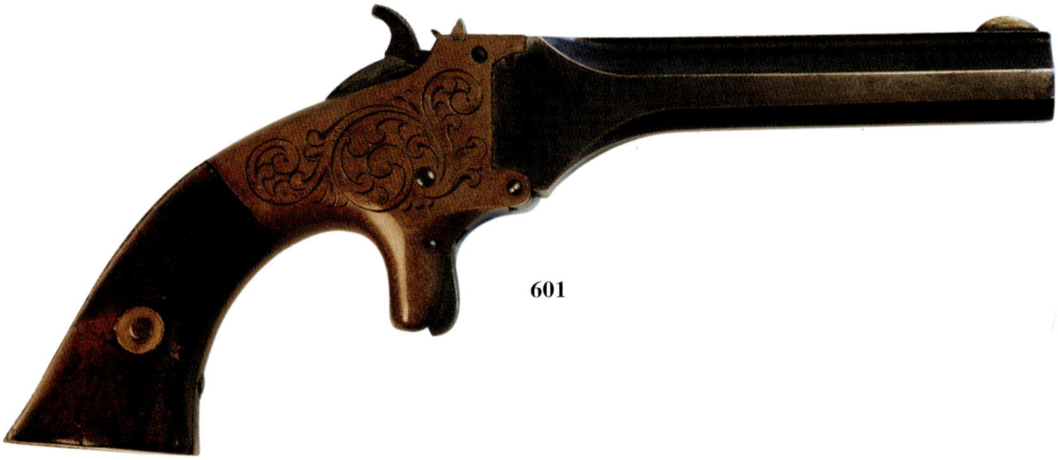

601

T.J. Stafford operated in New Haven, Connecticut. The piece illustrated is marked PATENT APPLIED FOR. Later examples are marked PATENTED MARCH 19 1860.

Nearly all pistols of this maker have some engraving. The coverage may be confined to frame; several observed also have engraving on barrel. Finish of silver, gold wash, or combination of the two. No nickel examples have been noted.

Foliate C-scroll engraving was standard form of decoration.

Backstrap **1863**

Backstrap **13**

601

1753

Note stippled background and more complex frame design.

1863

121+ Tipping & Lawden 121+

468

This firm located in Birmingham, England produced pepperboxes under license from Christian Sharps of Philadelphia. This and the Sharps are basically the same. There were slight modifications of design throughout period of production. The pistol was produced from 1859 to 1879. Most were engraved; most came in compartmentalized wooden cases with accessories.

1018 1017 468 1631

The pieces were sent to engravers of the area; when engraving was completed, returned to Tipping & Lawden for final finish. Frank Sellers ("Sharps Firearms") provides a section solely related to marks of the engravers used by the firm.

Other patterns exist that are not illustrated. Some had gold inlays and ornate ivory or metal grips. The standard grip was figured gutta percha. Frames were usually silver plated, and barrels were blued.

This and the following page illustrate various aspects of serial number **1631**. To date, this is the only example observed with the entwined rope motif on sides of frame and standing breech. The backstrap pattern also is quite different from the norm. Note overlapping leaves on crosspiece of buttstrap.

468

1017

1018

1018

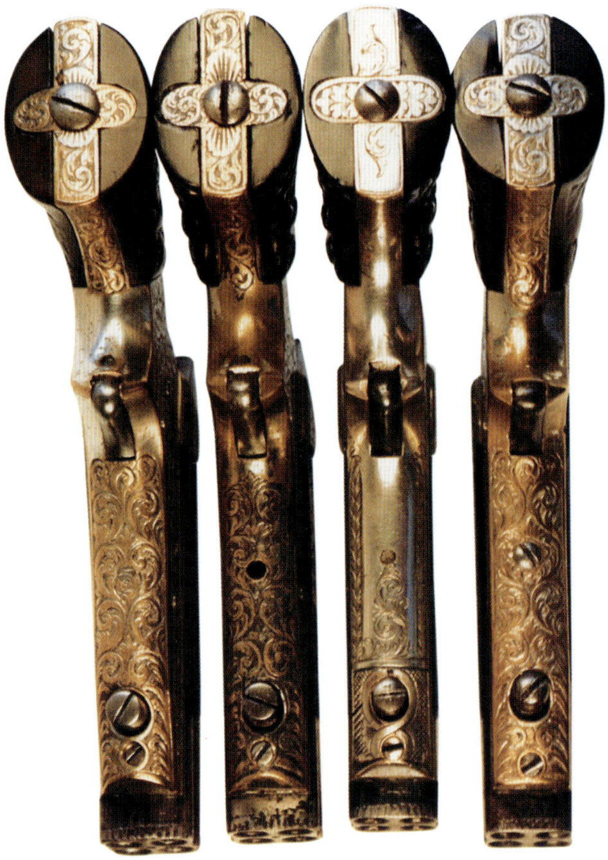

468 **1631** **1017** **1018**

Comparative sizes of .22 and .32 calibre models.

468

328+ Tranter 328+

89/11038

William Tranter was a prominent designer and producer of firearms. He was located in Birmingham, England. Best known for percussion revolver systems; production of cartridge models began in the early 1860s.

Guns were serialized in regular pattern, but there also appeared another number on top left side of frame forward of cylinder. That was not a serial or patent number, but probably a retailer's inventory number; it was stamped differently than rest of inscription. Illustrations are labeled with both serial and inventory (?) numbers. Often the name and address of a retailer will be found engraved on barrel.

This was a finely made gun, and the quality of engraving was equal to it. Most are fully engraved, including at times the backstrap and buttcap. Other than slight line decoration near muzzle, the barrels and cylinders were not engraved. One rather plain piece is illustrated herein - the exception rather than rule.

Most examples observed are housed in well made cases similar to those associated with the Tipping & Lawden pepperbox (see 121). They would have been quite expensive, obviously made for the higher class trade.

The engraving pattern is quite consistent. One feature nearly always appearing on the gun was a rosette surrounding the hammer pivot screw. Note **98/11184**; this and periphery border lines are the only engravings found on the piece.

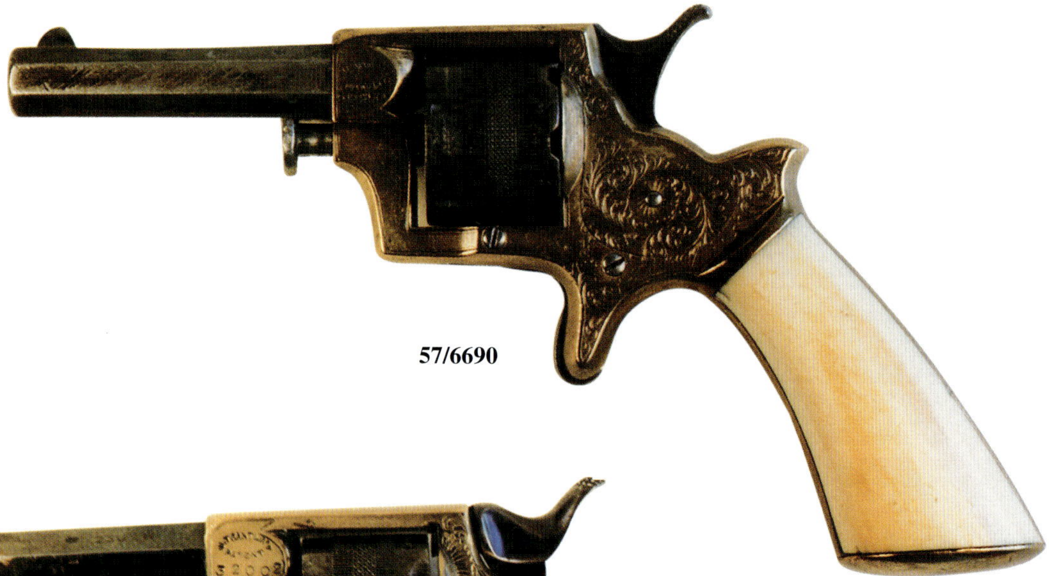

57/6690

79/32002

98/11184

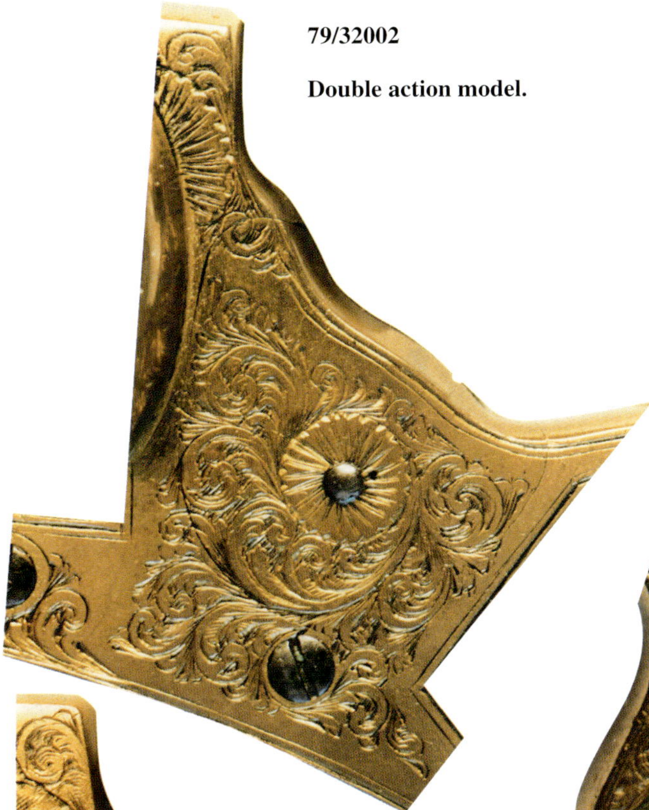

79/32002

Double action model.

57/6690

Single action model.

89/11038

Single action model.

Buttcap

79/32002

57/6690. Note monogram on backstrap. This may have been the personal gun of J.H. Crane, gun dealer. Monogram is the same as trademark on label from pistol case which houses an English Smith & Wesson copy revolver.

Buttcap pictured in upper right corner.

584+ Whitneyville Engraving Patterns 584+

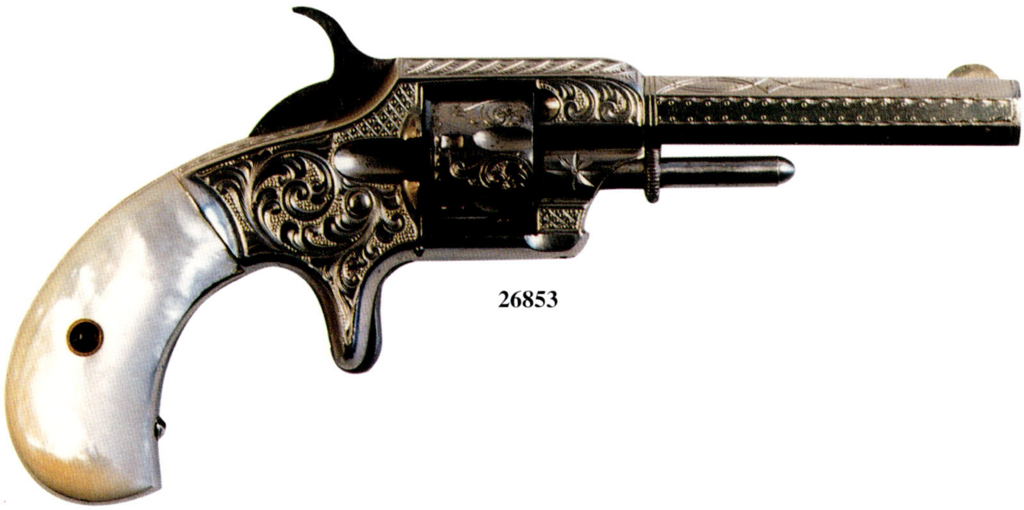

26853

Engraved pieces appear quite frequently. To date, all types with the exception of Type Three (D586, marked UNION) have been observed. Several of the patterns are found on different Whitneyville types.

When a revolver is engraved, all major parts (from sides, topstrap, backstrap, barrel and cylinder) have some type decoration. No partially engraved pieces have been seen, as has been observed on revolvers of other makers such as Manhattan, Deringer and Rupertus.

No engraved pieces have been seen with a blued finish. Some have silver but most have nickel plating. No gold wash has been seen.

Whether the engraving was done in plant or contracted with outside craftsmen is unknown. Also not known is whether one or a few people executed the engraving over the total period of production. The widely varied patterns may be changes made by one group of engravers over the years - or different people at different times.

Some patterns appear on the same type, then may appear on other types. There seems to have been no set pattern at any one time. The individual engraver apparently had complete freedom relative to patterns employed. There was no set style per type. As time passed, the quality of engraving became more shallow and poorer than preceding types.

The quality of engraving is average for the period. A few pieces have better quality engraving than the norm, may have been a "better" grade commanding a higher price. The Type 6 (D589) marked MONITOR definitely has an inferior grade engraving compared to those bearing the Whitneyville inscription. All lines are cut, not stamped. Exception: punched dots, small circles, and some wavy lines. Patterns on sides consistently have a stippled background. The patterns utilized are similar to patterns used on engraved examples of contemporary makers. No pattern unique to Whitneyville revolvers observed.

Pattern # + DESCRIPTION (Top, Center, Bottom)	LOCATION	TYPE	SERIAL #
1 Two lines circling barrel	muzzle	1 1 4	26853 22368 14515D
2 Elp-dmd-elp-1/2dmd	top + side flats	2	10854R
3 Elp-dmd-elp-dmd-elp-1/2dmd double lines at rear	sides other patterns other flats	1	22368
4 Elp-dmd-elp-dmd-1/2dmd double lines at rear	ud flats other patterns other flats	1	26853
5 Elp-elp-dmd-elp-1/2dmd double lines with dots outside lines	ud flats + sides	4 4	14545D 16633D
6 Elp-dmd-elp-dmd-elp-1/2dmd double lines with dots	sides other patterns other flats	5	13479D
7 Elp-dmd-elp-1/2dmd double lines with dots	ud flats other patterns other flats	5	2406A
8 Elp-elp-dmd-elp-1/2dmd double lines with dots inside + outside lines	ud flats other patterns other flats	5	112A
9 Scalloped edging; straight border Circles between scallops.	sides other patterns other flats	1	26853

elp = ellipse
dmd = diamond
1/2dmd = half diamond
top flat = top barrel flatside
ud flat = upper diagonal barrel flat
side = side barrel flat

TYPE 1 = d584
TYPE 2 = D585
TYPE 3 = D586 UNION
TYPE 4 = D587
TYPE 5 = D588
TYPE 6 = D589 MONITOR

Pattern # + DESCRIPTION (Top, Center, Bottom)	LOCATION	TYPE	SERIAL #
10 Extended leaves + spray	top, each side of MONITOR spray in front of MONITOR	6	11600
11 Rectangles slanted rearward. Scalloped tops. Circles + lines within each.	sides other patterns other flats	5 5 5	2406A
12 Clipped corner squares. Circle within each square	sides other patterns other flats	5	9375D
13 Acanthus leaves. Dot in center of each.	sides other patterns other flats	5	112A
14 Wavy line, single or double. Semi-circular cuts above and below waves.	ud flats other patterns other flats	5 5	12 10795D
15 Single wavy line. Alternating solid dots and curved cuts on each side of wave.	ud flats other patterns other flats	5	9375D
16 Rectangles, tops slanting forward. V cuts at tops. Circle + line in each.	sides other patterns other flats	5	10795D
17 Rectangles slanted rearward. Rounded tops. Series of dots within each.	sides other patterns other flats	5	12
18 Extended shaded leaves.	ud flats other patterns other flats	1	22368
19 Double wavy line. Circles above & below waves	ud flats other patterns other flats	5	13479D
20 Double line scallops both sides; spray at front. Single scalloped line. Short foliate clusters.	top muzzle ud + side flats, near rear	1	14320

Pattern # + DESCRIPTION (Top, Center, Bottom)	TYPE	SERIAL #
30 Foliate C-scroll. No perphery line. Deep shading. Lesser area coverage than on other patterns.	2 4	10854R 14515D
31 Foliate double C-scroll. Periphery line. Light shading.	1	22368
32 Foliate C-scroll. Pointed leaf at top front and into sheath. Periphery line. Light shading.	1 5 5 5 5	26853 10795D 12 9375D 13479D
33 Foliate C-scroll. Rounded scroll at top front. Medium shading. Periphery line.	5 5	2406A 112A
34 Foliate C-scroll. Double periphery line. Light shading. Shallow. Not as well executed as other patterns.	6	11600
35 Small foliate C-scrolls; no main scroll. More background stippling than engraving. Periphery line. Much deeper cuts than normal. Series of opposing scallops on top and bottom straps.	1	14320

The corresponding areas on right side of frame carry through with same basic patterns.

TYPE 1
14320

side	**35**
cylinder	**68**
forward	**96**
top	**50**

TYPE 5

	12
side	**32**
cylinder	**64**
forward	**94**
top	**45**

TYPE 5
9375D

side	**32**
cylinder	**66**
forward	**94**
top	**47**

TYPE 4
14515D

side	30
cylinder	63
forward	93
top	44

TYPE 2
10854R

side	30
cylinder	61
forward	92
top	42

TYPE 1 26853

Side	**32**
Cylinder	**60**
Forward	**90**
Top	**40**

TYPE 1 22368

Side	**31**
Cylinder	**60**
Forward	**91**
Top	**41**

Pattern # + DESCRIPTION (Top, Center, Bottom)		TYPE	SERIAL #
40	Series of clipped knife points, each with shaded blades. Grooved topstrap, pattern on each side of groove.	1	26853
41	Eight five pointed stars. Grooved topstrap, pattern on each side of groove.	1	22368
42	Plain, no engraving on sides of groove.	2	10854R
43	Stamped wavy line, small circles stamped between waves. Grooved topstrap, pattern on each side of groove.	4	16633D
44	Stamped wavy line, engraved curves between waves. Grooved topstrap, pattern on each side of groove.	4	14515D
45	Series of inverted Vs or sloped petals. Shading near apexes. Flat topstrap. Small circle within each petal on #12.	5 5	12 2406A
46	Crosshatched diagonal squares, circle in each square. Double wavy periphery line. Flat topstrap.	5 5	10795D 13479D
47	Crosshatched diagonal squares. Plain centers, stippled areas outside squares. Flat topstrap.	5 5	9375D
48	Elliptical pattern bordered by scalloped lines, dot under each scallop. Flat topstrap.	5	12A
49	Spray, pointing toward rear. Wavy line each side. Rounded contour on topstrap.	6	11600
50	Double lines of opposing scallops on each side of sight groove. Series of clipped knife points on each diagonal flat, each with shaded blades.	1	14320

Pattern # + DESCRIPTION (Top, Center, Bottom)	CYLINDER	TYPE	SERIAL #
60 Foliate c-scroll. Series of small triangles at rear.	fluted, short fluted, short	1 1	26853 22368
61 Double line of Vs, with curved arms pointing forward.	smooth, short	2	10854R
62 Foliate c-scroll, pointing rearward. No decoration at rear.	fluted, short fluted, long fluted, long	4 5 5	16633D 112A 2406A
63 Knurled oval toward front, 7-line spray extends rearward from oval.	fluted, short	4	14515D
64 Double line forward of locking slots. Foliate c-scroll, pointing forward.	fluted, long fluted, long	5 5	13479D 12
65 Double line forward of locking slots. Series of Vs, or chevrons, tapering in size toward front.	fluted, long	5	10795D
66 Double line forward of locking slots. Nine-line spray extends forward.	fluted, long	5	9375D
67 Single line with shading forward of locking slots. 12-curved-line spray extends forward.	smooth, long	5	11600
68 Spray resting on rear edge extends forward. Four or five branches in each spray.	fluted, short	1	14320

Pattern # + DESCRIPTION (Top, Center, Bottom)		TYPE	SERIAL #
70	Foliate C-scrolls, inscription space; elp-dmd-elp-elp. Stippled background in scroll area. Scalloped periphery lines with small circles.	1	26853
71	Spray pointing rearward. Double line elp-dmd-elp.	1	22368
71A	Spray pointing forward. Elp-dmd-elp-elp.	4	14515D
72	Small foliate with extended leaf directly behind hammer.	2	10854R
73	Spray, diagonal squares, circle in each square. Elp-dmd-elp-elp. Dots outside geometric segments.	4	16633D
74	Diagonal plain squares, stippled areas by squares. Double line elliptical loop.	5	10795D
75	Foliate C-scrolls. Double line elp-dmd-elp-elp. May or may not have background stippling.	5 5	2406A 112A
76	Shell + scroll behind hammer, series of chevrons, two small inverted Vs.	5	9375D
77	Foliate C-scrolls behind hammer. Scalloped double border lines ending in elliptical loop.	5	12
78	Series of diamonds, each with circle in center. Double periphery border lines terminating in double line elliptical loop. Vertical shading lines between diamonds and border.	5	13479D
79	Series of diamonds, each with small circle in center. No background lines. Single periphery border line terminating in V.	6	11600
80	Large sunburst near top. No periphery lines. Line of scallops extends upward and downward from sunburst.	1	14320

TYPE 5
2406A
75

TYPE 5
12
77

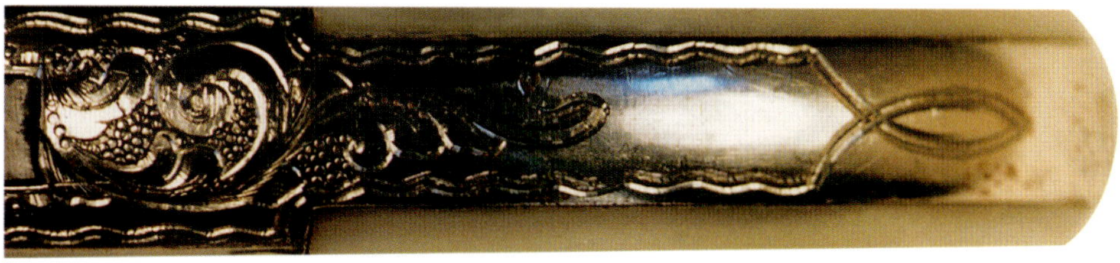

TYPE 1
22368
71

TYPE 4
14515D
71A

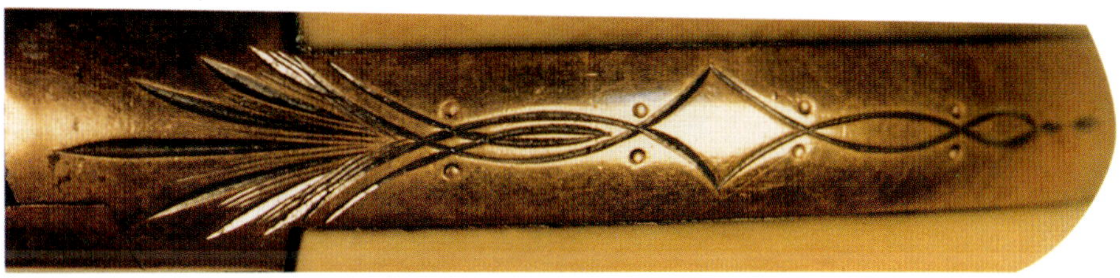

TYPE 4
16633
73

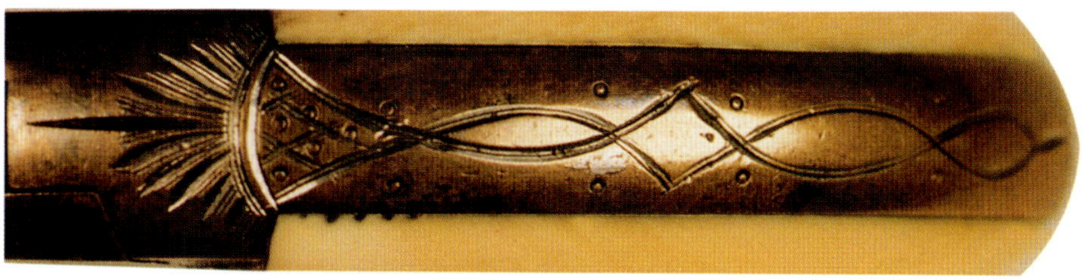

TYPE 1
26853
70

TYPE 5
10854R
72

TYPE 5
13479D
78

TYPE 5
9375D
76

TYPE 6
11600
79

TYPE 5
10795D
74

TYPE 1
14320
80

TYPE 5
12
side 17
diagonal 14

TYPE 5
13479d
side 6
diagonal 19

TYPE 1
14320
side 20
diagonal 20

TYPE 1
26853
side 9
diagonal 4
muzzle 1

TYPE 1
22368
side 3
diagonal 18
muzzle 1

TYPE 2
10854r
side
diagonal

TYPE 5
112A
side 13
diagonal 8

TYPE 4
16633D
side 5
diagonal 5

TYPE 5
9375D
side 12
diagonal 15

TYPE 5
10795D
side 16
diagonal 14

TYPE 5
2406A
side 11
diagonal 7

TYPE 1
26853
40

TYPE 5
12
45

TYPE 5
9375D
47

TYPE 6
11600
49

Pattern # + DESCRIPTION (Top, Center, Bottom)	TYPE	SERIAL #
90 C-scrolls; stippled background. Eight-pointed sunburst. Quilted diamonds, each with circle in center.	1	26853
91 Foliate. Foliate. Plain. (Plain grooves between sections.)	1	22368
92 Three leaves with stippled border. Plain. Plain. (Plain grooves between sections.)	2	10854R
93 Eight-pointed sunburst. Plain.	4 4	16633D 14515D
94 Quilted diamonds, each with circle in center. Plain. Plain.	5 5 5 5 5 6	12 2406A 9375D 10795D 13479D 11600
95 Quilted diamonds, each with circle in center. Plain. Leaf pattern.	5	112A
96 Foliate C-scrolls; stippled background. Spray. Plain. Series of scallops in grooves between sections.	1	14320

TYPE 5
13479D
46

TYPE 5
10795D
46

TYPE 5
2406A
45

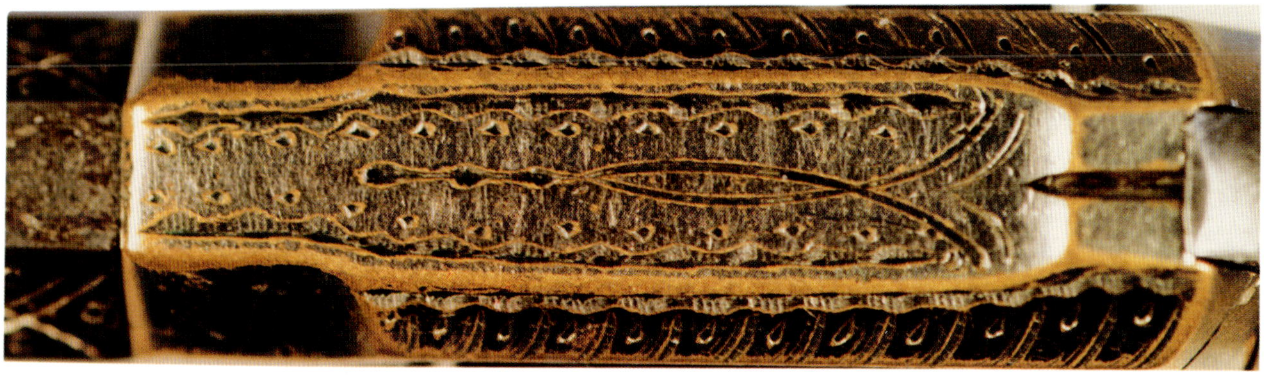

TYPE 5
112A
48

Bibliography

BOWEN, Taylor G. *James Reid And His Catskill Knuckledusters*. Andrew Mowbray Inc., Lincoln, RI, 1989. LOC 89-063243 ISBN 0-917218-37-X

BROPHY, William S. *Marlin Firearms*. Stackpole Books, Harrisburg, PA, 1989. ISBN 0-8117-0877-2

COPE, Kenneth L. *Stevens Pistols and Pocket Rifles*. Museum Restoration Service, Ottawa, Canada, 1971. ISBN 0-919316-64-6

DUNLAP, H. Jack. *American British and Continental Pepperbox Firearms*
Recorder-Sunset Press, San Francisco, CA, 1964
LCC 64-24121

EBERHART, L.D. & Wilson, R.L. *The Deringer in America, Volume II*. Andrew Mowbray Inc., Lincoln, R.I. 1993. LCC 85-60923 ISBN 0-197218-57-4

JINKS, Roy G. *Artistry in Arms, The Guns of Smith and Wesso*n. Smith & Wesson publication, Springfield, MA.

LAIDACKER, John. *Early Cartridge Handguns, .22 Calibre*. Precision Printers, Millville, PA, 1994. LCC 93-91773

SELLERS, Frank M. *The William M. Locke Collection*. The Antique Armory Inc., East Point, GA, 1971. LCC 73-91748

SELLERS, Frank M. *Sharps Firearms*. Beinfeld Publishing Inc., North Hollywood, CA, 1978. ISBN 0-9608122-0-2

WILSON, R.L. *L.D. Nimschke - Firearms Engraver*. John J. Malloy, Teaneck, NJ, 1965.

WILSON, R.L. *Samuel Colt Presents*. Wadsworth Atheneum, Hartford, CT, 1961.